Organic Fertilizers – History, Production and Applications

Edited by Marcelo Larramendy and Sonia Soloneski

Published in London, United Kingdom

IntechOpen

Supporting open minds since 2005

Organic Fertilizers - History, Production and Applications
http://dx.doi.org/10.5772/intechopen.77847
Edited by Marcelo Larramendy and Sonia Soloneski

Contributors
Ladislav Menšík, Lukáš Hlisnikovský, Eva Kunzová, Pham Toan, Ngo Duc Minh, Jozef Visser, Pawlin Vasanthi Joseph, Dao Van Thong, Niladri Paul, Utpal Giri, Gourab Roy, Kamila Kydralieva, Sharipa Jorobekova, Gaius Eudoxie, Micah Martin

First published in London, United Kingdom, 2019 by IntechOpen
IntechOpen is the global imprint of INTECHOPEN LIMITED, registered in England and Wales, registration number: 11086078, 7th floor, 10 Lower Thames Street, London, EC3R 6AF, United Kingdom
Printed in Croatia

British Library Cataloguing-in-Publication Data
A catalogue record for this book is available from the British Library

Additional hard and PDF copies can be obtained from orders@intechopen.com

Organic Fertilizers - History, Production and Applications
Edited by Marcelo Larramendy and Sonia Soloneski
p. cm.
Print ISBN 978-1-78985-147-2
Online ISBN 978-1-78985-148-9
eBook (PDF) ISBN 978-1-78985-089-5

For EU product safety concerns:
IN TECH d.o.o., Prolaz Marije Krucifikse Kozulić 3, 51000 Rijeka, Croatia, info@intechopen.com or visit our website at intechopen.com.

Meet the editors

Marcelo L. Larramendy, Ph.D., serves as a Professor of molecular cell biology at the School of Natural Sciences and Museum (National University of La Plata, Argentina). He is an appointed Senior Researcher of the National Scientific and Technological Research Council of Argentina and a former member of the Executive Committee of the Latin American Association of Environmental Mutagenesis, Teratogenesis and Carcinogenesis. He is the author of more than 450 contributions, including scientific publications, research communications and conferences worldwide. He is also the recipient of several national and international awards. Prof. Larramendy is a regular lecturer at the international A. Hollaender Courses organized by the IAEMS and former guest scientist at NIH (USA) and University of Helsinki (Finland). He is an expert in genetic toxicology and has been referee for more than 20 international scientific journals. He was a member of the International Panel of Experts at the International Agency for Research on Cancer (IARC, WHO, Lyon, France) in 2015 for the evaluation of DDT, 2,4-D and Lindane. Presently, Prof. Dr. Larramendy is Head of the Laboratory of Molecular Cytogenetics and Genotoxicology at the UNLP.

Sonia Soloneski has her PhD in natural sciences and is a Professor Assistant of molecular cell biology at the Faculty of Natural Sciences and Museum of La Plata, National University of La Plata, Argentina. She became a member of the National Scientific and Technological Research Council (CONICET) of Argentina in the genetic toxicology field. Presently, she is a member of the Latin American Association of Environmental Mutagenesis, Teratogenesis and Carcinogenesis (ALAMCTA), the Argentinean Society of Toxicology (ATA) and the Argentinean Society of Genetics (SAG). She has authored more than 350 scientific publications in the field, including scientific publications in research papers, reviewed journals and conferences worldwide. She is a referee for issues related to genetic toxicology, mutagenesis and ecotoxicology.

Contents

Preface

Years ago, the Food and Agriculture Organization of the United Nations predicted that the world population will be over 9.1 billion people by the middle of the 21st century. Accordingly, food production will have to rise about 70% above current levels to maintain pace with demand. One plausible method for obtaining this enhancement in food production would be to increase the amount of land available for agriculture. However, the conversion of natural forests and/or other wild habitats engenders a number of well-known negative impacts on climate change and global bio-diversity. Furthermore, it is accepted worldwide that such an expansion of agriculture could be responsible for approximately 12% of global warming. Regardless of its implications, sustainable agriculture must be based on providing optimal growing conditions for plants in order to achieve optimal crop production from the land over a season. To not only optimize crop yield but also to reduce the negative impacts that agriculture can exert on the environment, it is mandatory that farmers adopt the best agricultural practices. Agriculture in the 21st century faces several challenges, including: producing meat without raising animals, better irrigation management for agricultural processes, the development of genetic engineering for drought-tolerant and higher-yielding crops, the improvement of agricultural precision and aquaculture, the sustainable development of biofuels, and the promotion of organic agriculture around the world. However, intensifying food production must be achieved in an environmentally safe manner through ecological intensification to increase the yield per unit of land, approaching the maximum available yield of farming systems, with minimal or no negative environmental impact. It is evident, then, that fertilizer selection, as well as its rational use, is key to meeting this challenge.

Perhaps the most important of the major objectives of farmers, members of National Administrations, and the suppliers of agricultural inputs is to both stimulate the use of appropriate agricultural practices and to guarantee the availability of suitable fertilizers in the market. Techniques such as crop rotation, minimum tillage, and crops grown under cover tend to maintain the structure and quality of soils. The correct selection and application of fertilizers is directly determined by the correct dose, the right place , and the right time to use the product.

By definition, a fertilizer is the name given to any material, either of natural or synthetic origin, that is applied to soils or to plant tissues to supply at least one, but often more, of the nutrients essential for plant growth. The majority of fertilizers employed in commercial farming provide the three main soil fertilizers (namely, nitrogen, phosphorus, and potash). These fertilizers are extracted from minerals (e.g., from phosphate rock) or produced industrially (e.g., ammonia). In contrast, the other type of product employed is the organic fertilizers, which are derived from animal matter, animal excreta (manure), human excreta, and vegetable matter (e.g., compost and crop residues). Naturally-occurring organic fertilizers include animal wastes from meat processing, peat, manure, slurry, and guano. Dependence on organic nutrient sources is a central characteristic of organic agriculture, which uses nutrients derived from sources such as livestock and green manure and even several types of compost to meet crop demands in intensive cereal production. One of the advantages of the use of organic fertilizers is that they provide crops with nutrients over a long period of time in a slow and extended release process. Accordingly, more research on improving efficiency and minimizing losses from organic natural resources is needed to determine their costs and benefits, and to develop optimal agricultural practices to avoid the use of synthetic inorganic fertilizers.

This book, *Organic Fertilizers – History, Production and Applications*, aims to provide an update on research issues related to organic fertilizers, highlighting their importance in sustainable agriculture and the environment. We aimed to compile information from diverse sources into a single volume and to give some real-life examples, extending the appreciation of organic fertilizers that may stimulate new research ideas and trends in relevant fields.

This book comprises of seven general chapters describing the history and production of organic fertilizers, including several manure types and other farming-derivative products, and the advantages of employing organic rather than mineral fertilizers. The first chapter comprises an extensive and detailed review describing the past and current status of the various organic fertilizer sources and how these fertilizers have been employed throughout history, depicting their strengths and drawbacks. The second chapter aims to provide information about how organic matter and nutrients play an important role in terrestrial ecosystems and agro-ecosystems. This chapter also describes how long-term application of mineral fertilizers and farmyard manure maintains the health of soils. The third chapter reviews the importance of composting, a very old methodology, summarizing some of its basic principles, which have been appreciated and employed in crop production for centuries. This chapter describes the rapid progress achieved during recent years through scientific studies of the underlying biological and chemical processes involved in composting. Furthermore, the chapter points out how some of these studies have served to clarify several factors that can act to produce a finished compost that is both valuable to agriculture and relatively safe from the viewpoint of public health. The fourth chapter provides information about employing the products of anaerobic fermentation of agricultural wastes, produced by a consortium of methanogenic microorganisms including humic-like substances, for applications such as plant growth biostimulants, organic-mineral fertilizers, and phytohormones. The fifth chapter depicts the importance of compost teas as organic fertilizers, describing the nature and behaviour of these nutrient sources, and provides an analysis of their effects and mechanisms, stressing how compost tea represents an ideal beneficial product in any cropping system. The sixth chapter aims to provide information about the role of vermicompost, a pollution-free and cost-effective product, employed in many applications to increase water-holding capacity, crop growth, and yield, and to improve the physical, chemical, and biological properties of the soil to increase the production of plant growth regulators. Lastly, this book includes a final chapter discussing and highlighting the improvements in crop production that have taken place during the last 30 years in a country from the South East of Asia, Vietnam. The authors emphasize that Vietnam already produces organic fertilizer from a range of materials using different production technologies, but production capacity is small and does not meet the demands of organic agriculture.

Finally, as previously indicated in a book we published some years ago entitled "Organic Fertilizers – From Basic Concepts to Applied Outcomes", it seems more than evident "that future agricultural practices will irreversibly shape the Earth's land surface, including its species, geochemistry, and disponibility of surface to the people living on it. We hope that the information presented in this book will be of value to those directly engaged in the handling and use of organic fertilizers, and that this book will continue to meet the expectations and needs of all those interested in the different aspects of the use of organic fertilizers to achieve a sustainable agriculture without compromising environmental integrity".

The contributions made by the specialists in this field of research are gratefully acknowledged. The publication of this book is of great importance for those researchers, scientists, engineers, teachers, graduate students, agricultural agronomists, farmers, and crop producers who can use these different investigations to understand the advantages of the use of organic fertilizers.

Marcelo L. Larramendy Ph.D. and Sonia Soloneski Ph.D.
School of Natural Sciences and Museum,
National University of La Plata,
La Plata, Argentina

Chapter 1

Opening History: Gaining Perspectives

Jozef Visser

Abstract

After Second World War, historical agricultural systems that gave pivotal roles to organics were effectively locked away, with a warning on the door 'Liebig disproved it all!'. The recent digitalisation of a vast amount of historical literature gave us the key to unlock the door. It opens not to a dusty archive but to a land with great treasures. Entering it we regain a perspective on the pivotal roles of organics in agriculture but not without effort. We lost contact with the soil when after Second World War, we denied farmers' practices and focussed at fertiliser industry instead. Proud of our construct, 'modern agriculture', we nevertheless positioned the statue of Liebig the frightening warrior in front. It is not easy to get rid of a mix of pride and fear. Still, historical evaluation helps us to uncover what was hidden and equips us to rediscover the roles of ever-local organics as administered by local farmers.

Keywords: Liebig, peer review, soil quality, mixotrophy, mineral solubilisation, De Saussure, post-war policy, legumes, Olsen P test, extended N- and P-cycles, modelling

1. Introduction[1]

The eighteenth and nineteenth centuries were very rich in agricultural literature, and very much of it is relevant as to the subject of the present volume. Yet, nearly all of this literature has been consistently neglected after Second World War. No doubt the reader knows the one-liner 'Liebig disproved Thaer's humus theory of plant nutrition and proved mineral nutrition instead, then with the introduction of industrial fertilisers crop yields could grow steeply'. Now although one-liners cannot make up for history, they still can induce the neglect of historical sources, and exactly that happened when the Liebig one-liner was used to sideline the roles of organics in agriculture and plant nutrition. Historian Frank Uekötter (now in Birmingham) took a close look at the period between the World Wars and after Second World War and showed that as to agriculture and soil, an all-out 'knowledge erosion' occurred (See his 500+ pp. 2010/2012 *The truth is in the field*—in German). When we go further back in history, it becomes clear that Liebig himself was at the roots of this 'knowledge erosion'.

[1] The present contribution embodies original historical research so there are no reviews yet, and we have to list the original sources. The historiographic approach used here has its roots in the work of the prominent science historian Reyer Hooykaas († 1993), while the recent Leiden PhD theses of Karstens [162] and Bouterse [163] give expositions of its 'evaluative historiography'. See for Uekötter's free-access publications on the loss of history in agronomics [164–169].

2. Reconsidering the Liebig thesis

We know from his correspondence with the leading chemist Berzelius that Liebig started his first experiments in crops growing only in 1841, a year after he published his *Chemistry in its application to agriculture and physiology*. Berzelius asked him the details of the experiments but never received them [1]. Yet, in the 1850s Liebig prescribed the setup of experiments and their analysis in the Bavarian agricultural experiment station. At the core of it was, *first*, his equation of 'elements' and 'nutrients' and, *second*, his denial of mineral-organic interactions. The first was a misnomer also in 1840 but was backed up by the 'element juggling' that was at the core of Liebig's plant physiology and physiology at large (remember he denied enzymatic catalysis and catalysis at large). The second was at variance with a very long time of agricultural experience. Yet, in combination they suggested that something completely transparent—the nutrient element construct—now could guide agriculture in its dealings with soils and crops, and it was this that was difficult to resist in these high days of rationalism (on this see [2]).

Note also that the search for a 'fertilising principle' had a long history already. With saltpetre known for centuries, its quick action in plant growth stimulation was discovered early and brought some authors to its identification with the 'fertilising principle'. Early in the nineteenth century, there had been some regional efforts in France as well as in England to use it in agriculture, but these had soon been abandoned because of lodging and plant disease problems. Most farmers were at pains to build soil quality (see later), aware that there were no 'fertilising principles' offering a short cut. Once soil quality was achieved, some kind of 'fermentation' was thought to have a role in mobilising plant nutrients.

But then in the middle decades of the nineteenth century, interregional and international transport grew strongly, and especially with guano supply, the old dream of fertilising principles seemed to materialise. It was in these decades that the 'clearness' of the Liebig concepts and methods stood out against the high complexity of soil and soil organics. The picture was seductively clear: analyse your soil for abundant and scarce 'nutrient elements', analyse your crops for those elements removed, resupply them in your mineral fertiliser, and the problem of 'soil exhaustion' is no more. The only role of organics was that of suppliers of mineral nutrients and carbon dioxide on decomposition.

Still, common experience first of all in vegetable gardens had established the value of soil organics and of the extract of garden soil to crop growth. And so the *Société centrale d'agriculture de Rouen* at the end of the 1840s came with a price contest about the subject. Soubeiran won the contest with his essay 'Chemical analysis of humus and role of manure in plant nutrition' [3, 4]. In accordance with De Saussure 1841, he proved that dilute ammonia dissolved an important part of the soil organics and stressed that ammonium carbonate deriving from disintegrating manures had the same effect. This solubilised part of 'humus' then entered the plant. The fact that the plant derives most of its carbon from the atmosphere was not to deny the importance of humus uptake, 'for if the absorbed humus gives effectively a nutrition that increases the plant's vitality, and that causes the number and volume of absorbing organs to increase, the plant will derive much more from the atmosphere. The humus, without having provided all of the carbon, will nevertheless be the effective cause of the abundant production of wood and other parts of the plant'.

This plant growth promotion aspect of humic extracts from fertile soil was accepted also by researchers who doubted true assimilation of humics. Soubeiran made careful experiments, emphasised the many valuable functions of humus, and gave close consideration of manures and their interaction with humus [5]. Malaguti, the leading French chemist known also for his work on agricultural chemistry, followed on the Soubeiran experiment and used the balance to prove the humus uptake [6]. Then in 1862 Corenwinder [7] showed carbon dioxide uptake by roots to be unimportant (as a rule), thus disproving Liebig's explanation for plant growth stimulation by humics. So research into the connections between crop growth and soil organics continued, with Eugene Risler from Switzerland soon followed by Pierre-Paul Dehérain in France and Ewald Wollny in Germany.

Yet, for some time the Liebig approach stayed dominant for it had been 'institutionalised' in most of the agricultural experiment stations, with their directors often from the Liebig school and quite uncritical to the Liebig doctrines. E. Wolff's *The natural law foundations of agriculture* [8] offers us an example. Jacob Johnson, a leading agronomist from (then) Russia, compared it with historical and experimental reality and found it wanting on both points [9]. When Wolff from the possibility to grow certain plants in mineral nutrient solution 'proved' the non-uptake of humics, Johnson remarked: 'So because we do not provide them with humics, therefore they cannot take it up and assimilate it! No doubt that is right; a man also cannot eat bread and digest it when he has none!'.

Leading scientists like Johnson [9, 10] and Risler [11, 12] reminded Liebig and his followers of historical agronomy. Although they published in leading journals, they had little response. Liebig's followers made their own accounts, e.g. [13]. That author is very selective in his sources, gives citations that are one-sided (e.g. about Thaer), and misses out on most of the leading agronomic publications from the eighteenth and nineteenth centuries. In fact, such an account is worse than no account. But organic practices were still part and parcel of agriculture and also in agricultural experiment stations and soon field experiments brought renewed attention to the importance of organics (e.g. Dehérain's and Wollny's research). Toward the end of the century, we see the Liebig doctrines losing their grip on the minds, not unlike what happened with his concepts and methods in physiology. Still, if we look at the 1840s, Liebig's influence could have been of much shorter duration.

The agronomist of great standing Schmalz [14] in 1841 published his 'To Julius: An open letter to Justus von Liebig' that makes clear how complex the discipline is on which Liebig wanted to impose his 'science' [15]. This followed on Liebig's 1841 [16] cross 'rebuttal' of Carl Sprengel's review [17] of Liebig's 1840 book [18] that, in conformance with his earlier work on humics [19], started with emphasising the roles of humics in plant growth and nutrition. Note that Sprengel's authority as editor of the (Prussian) *General Agricultural Monthly* derived from his thorough acquaintance with both agriculture and soil chemistry, so people now asked 'Sprengel or Liebig?'. To bring the discussion at the required level, Schmalz wrote *Aphorisms from plant nutrition learning* [20] in which he focussed, also from own experiments with nettle, more specifically on crop nutrition. But Liebig did not respond.

In those same years, Liebig's denial [21] of the roles of organics in plant nutrition, as against Théodore De Saussure's proofs [22, 23], was questioned by leading biologists of the age (von Mohl, Fürnrohr, Schlechtendal). When Trinchinetti's [24] as well as Jacob Johnson's [25] experiments corroborated those of De Saussure, the matter was settled, and Schlechtendal wrote: 'So it seems that Sprengel's teaching ….is right, yet unfounded Liebig's thesis that humus provides the plants only carbonic acid' [26, 27]. Note that Liebig in the 5th edition of his book mended many of the botanical and other faults that Hugo von Mohl had listed in his 1843 *Dr Justus Liebig's relation to plant physiology* [28]. He even removed the account of forester Hartig's experiments—all experiments he had (De Saussure had indicated Hartig had injured the roots of his plants). But von Mohl's [28] primary objection that he did not use the means available to him as a chemist to study the subject experimentally did not move him to an experimental approach. von Mohl predicted [29] that also the 5th edition would lead many astray because 'it lacks any and all historical account'. Liebig kept silent about it all including the *peer review* conclusion (see also [30]), but note that Hugo von Mohl [31] maintained the final verdict in his famous *Principles of the anatomy and physiology of the vegetable cell*.

Still, we see Liebig later in the 1840s hesitating (in a letter to Wöhler, see [32]) about further involvement with agriculture. Leading authors like Petit-Lafitte [33, 34], Schubert [35], and Fresenius [36] still emphasised organic uptake by plants (note Remigius Fresenius was the leading authority in analytical chemistry). Note also that Liebig's best student Adolf Strecker in 1848 acquired the *Venia legendi* with his researches about the chemical constitution of oxen bile [37], with one of the theses accompanying the account [38] 'Organic substances are nutrients for plants'. By then Liebig's patented mineral fertiliser had proved to be a failure [39], factually disproving his 'minerals-only' doctrine. Moreover in 1847 the Prussian Agricultural Council initiated an interlaboratory round for the determination of mineral nutrients in soils in connection with plant uptake and found it outside the possibilities of chemistry [40]. *First*, the changes in question, as calculated from the total mineral contents of the crop plants covering the field divided by the field's arable soil volume, were within measurement error (they would say so). *Second*, the differences in determinations (of in-soil quantities) between laboratories were excessive, a problem that would haunt such determinations for the rest of the century (and later). The Liebig model of crop nutrition, though 'claire et distincte', proved chemically unworkable when confronted with real soils.

But the 1840s were a decade of famine and revolution, not a decade in which careful study and evaluation decided about the events. Moreover many scientists of the old guard died in those years—Berzelius, Schwerz, de Saussure, Schwann, and quite some others—and the transfer of their roles in 'peer review' to younger scientists proved complicated enough. Liebig managed to draw that role to himself in connection with laboratory chemistry and then used this authority to once more give his judgements about bordering disciplines. In due time his opinion of catalysis, fermentation, and many medical subjects proved mistaken, and by the end of the century, his contributions to these fields were mostly passed over quietly. But in such disciplines, his opinions were discussed besides those of others; they did not shape the field, and so in due time, Liebig's errors could show up (cp. Pasteur's experiment disproving Liebig's abiotic theory of fermentation).

As it was, after the lingering indicated, we see from 1850 on Liebig renewing his efforts to impose his 'science' on agriculture. His public fame depended on it, including his fame with the king of Bavaria who called him to Munich. Leading agronomists compared statements and generalisations of Liebig and his followers with real-life and historical agriculture, e.g. [9–12, 41–46]. It was to no avail because Liebig did not respond but constructed his division of 'agronomy before 1840 vs agronomy after it' instead (it found its way into the Liebig one-liner). And he reverted to his former cross approach, with 'fire and the sword' [47], replacing scientific discussions. When Mulder in 1865 published this three-volume overview of agriculture and agricultural chemistry [48]—the best of those decades according to van Bemmelen 1901 [49]—Liebig did not enter into discussion at all but instead wrote with biting sarcasm about Mulder. The way in which he in those same years managed to silence the very learned Fraas was still more infame. Fraas' *Book of Nature for Farmers* [50] for a broad public as well as his *The root life of crop plants and the increase of yields*' [51] had no equals in those years ([51] is still very profitable to read).

There is, in short, ample reason *not* to take Liebig as our guide. But high-level researchers in those years who were well-versed in both the agriculture of their age and in supporting disciplines we can take as our guide in restoring the history of agronomy in the second half of the nineteenth century including the subject of soils, organics, and crop nutrition. Now with the Liebig's stop sign removed, we of course start wondering about the wider history too. We will see that the farmers' knowledge of their soils can be our guide.

3. Soil quality as baseline

Estienne et al. [52] in the paragraph *About ground and soil of an estate* give a short list of characteristics of a fertile soil (S. 33): (1) A strong rain does not lead to mud; the ground drinks in the water and keeps it for a long time. (2) On a fertile soil, we meet a strong and dense vegetation (in the wild). (3) Its soil solution has a sweetish taste [see under]. (4) When we dig a hole, put the earth aside for 2 or 3 days, and then fill it again; a good soil will leave a small hill, a medium soil will just fill the hole, and a bad soil will leave a hollow. (5) With a good soil, the first rain after some drought in spring will give a pleasant odour. We find a similar list in [53–55]. Evidently also three or four centuries ago, farmers realised that—in our words—infiltration capacity and soil structure were characteristics of a good soil. The 'hole filling test' is part of the 'spade test' and fits in with other visual tests of our day [56–60]. The early authors took also good note of biological criteria, plant growth first of all. The 'sweetish taste' of the soil solution from fertile soils they considered connected with soil quality and plant nutrition—and indeed soil carbohydrates are central to both micro-aggregate formation and mineral solubilisation.

The same or a strongly similar list appears in other early books on agriculture or horticulture. Heresbach/Googe [61] adds still another biological indicator: birds following the ploughman to feed upon worms, etc. We find closely similar lists in [62–64], the most extensive being [65]. As to the era before the printing press, Verena Winiwarter has given us some admirable overviews [66, 67], from which we learn that such a soil quality approach was known already in antiquity and Middle Ages. A core element of sustainable agriculture [68] it is its guide also when seeking to upgrade soils. Farmers in the past evidently had a notion of 'good soil' (and how to build it)—or they would not have survived [2]. These practical standards of experienced farmers and gardeners as to good soil are a backbone of agronomy throughout history.

In the eighteenth century, 'earths mixing' became a chief means to upgrade soils, with farm manure being the 'soul' of the mixture ([69], p. 118). Marling and chalking became widely known but were two examples of this general practice of 'mineral fertilisation'. The aim of it all was soil quality ([70], p. 15), but of course the diverse 'earths' brought also their own minerals with them. Pastor/agronomist Mayer tells us this 'earth mixing' originated with Swiss farmers at the beginning of the eighteenth century, farmers who had to be very careful with their soils [69]. Lüders [71] is all about research, evaluation, and use of the 'kinds of earths'; it is also the chief subject of [72] on general agronomy, and [73] gave an account of nearly 300 soil probes from all over the kingdom of Hannover (research committed by the government).

Mineral and organic components belonged together in this approach, something that was strengthened still by the general concept of *nutritious matter for plants: it consists, next to water, in fine earthen particles, fatty components, and salts* [74]. Most other authors used 'oils' for 'fatty components' but the meaning of it is clear: the concept of plant nutrition of these eighteenth-century authors fits closely to our concept of mixotrophy. There were of course differences between the chemistry, etc. of those authors and ours, but when we read Home's [75] chemical analysis of different soils with the means then available we see that we should not exaggerate these differences. What we as people from the twenty-first century need on background information, we can learn from [76].

But then from the latter decades of the eighteenth to the first decades of the nineteenth century, chemistry saw fast developments, best known from the name of Lavoisier, and agricultural chemistry partook of this 'turbulence'. But note it was not a backwater: Hermbstädt who translated all of Lavoisier's works in German was also the editor of the *Archive of Agricultural Chemistry* (in German), the first journal of its kind. From about 1820 on a less turbulent period set in, mixotrophy became quite broadly accepted again, now in terms of the new chemistry and in combination with photosynthesis. Two examples will do.

The leading agronomist Zierl (1797–1844) from Bayern in his 1830 *Plant Production* textbook [77] emphasised—critically following Carl Sprengel—the roles of humics in *solubilising mineral soil constituents* for plant uptake. This followed upon his acknowledgement of photosynthesis and the extensive treatment of mineral and organic plant nutrients. Zierl considered plants to have important roles in soil nutrient cycling, legumes in rotations with their long roots bringing nutrients in reach again of other crops. In conformance with earlier agronomy, Zierl maintained a soil quality approach.

That is true also of Johann Nepomuk Schwerz (1759–1845) who acquired his encyclopaedic knowledge of the art of agriculture especially on his many journeys on foot through regions of Belgium and Germany. He was conversant also with the life and practices of the small farmer (Thaer decidedly not). In his 1823 *Manual of Practical Agriculture* [78], he gave a summary of the crop nutrition knowledge of those years: 'So when indeed the plants derive their chief nutrition from the atmosphere and besides also from the residues of vegetable and animal bodies: it is not to deny that also mineral bodies, under which we count first of all the chalk, contribute to the growth of vegetables, and not just in stimulating, solubilising, manure-mediating roles, but also as true nutrients'. This mixotrophic approach was essentially maintained in the 2nd edition of Schwerz' manual that was published in 1837 and no doubt was available to Liebig.

At mid-nineteenth century, the soil quality approach was upheld as we see from a wide range of textbooks (including [79]). Next there is ongoing development especially in the *Progress in Soil Physics* series (1872 f.) that was edited by Ewald Wollny who also included reviews of soil biology in the serial. Wollny further developed the concept of Bodengare, 'active soil' (e.g. [80]) from a soil science point of view. Others then continued, with Johannes Görbing the best known author [81] (see also [82, 86]) and scientists like Sekera further developing theory and practice after Second World War [83, 84].

Soil quality as the leading concept was upheld also when the great agronomist and plant breeder Fruwirth in the 1921 edition of his textbook [85] specified 'main fertilisers' as those that were building the soil biologically, physically, and chemically, with farm manures and green manures the standard and others and especially mineral fertilisers as 'additional' only. When cheap nitrogen fertiliser in

Germany after First World War was pushed by the industry-government complex to remedy the large drop in yields that was a consequence of the war, it did more harm than good [86]. Again classical agronomy with its soil quality approach came to the rescue and showed the way to restore the soils and so the yields. Likewise it was the work of the Soil Conservation Service in the USA along the lines of classical agronomy and soil science that brought soil deterioration to a halt. For the past few decades, similar 'organic' approaches proved necessary to restore structurally deteriorated soils in Germany and to revive desertifying regions in Mediterranean countries.

4. Loss of history and quality

A core element of this soil quality approach was and is the use of legumes in agriculture. Research in legume use between the World Wars was greatly advanced by Fred and co-workers in the USA, by Thornton and Nicol in the UK, and especially by Virtanen in Finland (references in [87, 88]). Legumes were quite central to agriculture in parts of Finland and the Baltic countries, with coculture of peas and oats being a prominent part [89, 90]. Johnson [89] already concluded that the oats thriving in the coculture received nutrition from pea root exudates. In big parts of Latin America, farmers used the milpa system of coculture for ages already [91, 92]. In the USA the common farmer used legume rotations without N-fertilisers; industrial N-fertilisers were used by big landowners in the US South.

From the point of view of soil quality-based agriculture (and therefore of sustainable food provision), fertiliser-only agriculture was the wrong choice, so when Virtanen saw the latter policy coming, he wrote some reviews (in 1953) emphasising the advantages of legume use. Importantly it costs the common farmer only her labour and no money, and there were many poor farmers in Finland before and after Second World War, as there are now the world over. But with the takeover of the USDA by the Republicans at the 1942 elections, agricultural policy got redirected from its focus at the common farmer under the New Deal to its focus at large-scale agriculture as practised by the big landowners of the US South. After the war industrial N-fertiliser was offered at low prices—process facilities had been financed by the government in connection with explosive production for war—and was put at centre place in the new policies. With classical agronomy very critical of such a change, the 'Liebig doctrines' were advanced in its defence, so we turn once more to their historical origins.

De Saussure in 1841 lectured at the 9th Scientific Congress of France—that chose him as its president—about the question: 'The ternary and quaternary organic matters can they – or not – be assimilated by the plants, after being absorbed by their roots?'. He used the results of his own recent experiments to give an answer to this question. His slightly edited lecture he published already before the end of the year in the scientific monthly of his home town Geneva: '*What role is there for ternary and quaternary compounds in plant nutrition, not just the few binary ones mentioned by Liebig, considering that soil organics as a complex mixture of compounds could contribute to plant nutrition?*. So De Saussure stated the problem at chemical compound level, departing from the conflation of 'element' and 'nutrient' that had been quite common up till then. His conclusions were as follows: (1) fertile terrains contain a mixture of soluble and (mostly) insoluble organic substances, and uptake of the first by plant roots forms a powerful addition to the nutrients it receives from air and water [CO_2, H_2O], (2) a slow fermentation of the insoluble organic substances renews soluble organics, and (3) most plants do not assimilate gaseous nitrogen and receive only little ammonia from the air, so nearly all the N they contain is from absorption of soluble organic substances.

This could have been the start of the development of a true soil-plant N-cycle because it used **(a)** a chemically meaningful concept of 'nutrient' **(b)** linked to the soil by the use of the dynamic concept of 'humus' that had been around a long time already (and that included dynamic interactions with minerals in soils). But then Liebig forced a rupture by **(a)** reverting to the element-as-nutrient parlance and so disabling chemical research **(b1)** denying the need for organic fertilisers: well-growing plants would of themselves leave ample organics in the soil (later dubbed 'self-fertilising plants') **(b2)** denying also the diverse direct roles of soil organics in crop growth, so legitimating plant nutrition studies disconnected from the soil and using a minerals-only approach (as in the work of Knop, Sachs, and later Hoagland).

As indicated, after Second World War, the new agricultural policy was presented in terms of these 'Liebig doctrines'. Melsted, for example [93], wrote *Since 1950 the corn belt farmer has been able to choose whether to grow or buy his nitrogen*. But legumes were recommended as green manures for several reasons: *first* as soil builders and strongly so after the Dust Bowl of the 1930s and *next* as natural resource 'slow release fertilisers'. American farmers practised rotations with legumes instead—with the exception of the big landowners of the American South who from 1942 on were redirecting agricultural policies (Jamie Whitten). Before that date mixed farming was strongly encouraged (USDA Yearbook of Agriculture 1940). Melsted offered the farmer nothing with which to make a choice but locked him in the 'P,K,N cage'.

Now humus management was upheld by the Soil Conservation Service under Hugh Bennett. But when Bennett retired in 1951, a 'straw man' was appointed in his place, and attention was diverted away from humus management, etc. A 'flight forward' was chosen to defend the change: the fertiliser-fed crops would surely leave enough organic matter in the soil! Classical teaching on humus management got dubbed 'lamentations' by Joffe [94], and he wrote that it came from '*agronomists who then dominated the field of soil fertility. ... A perusal of the writings of these specialists.... reveals the development of a soil organic matter mentality complex*'. In other words, Joffe and others who like him sounded the new agricultural policies simply denied the research of leading scientists in the field. When policy makers next lifted the unqualified adherence to the 'Liebig doctrines' to the status of 'civil obedience' (for advisors and researchers), the system got out of control for its lack of correction from the real world 'out there' that of ever-local farmers, plants, soils, and ecology. Post-war agronomy in countries like the USA, the UK, and the Netherlands was government-directed not just as to specific regulations but as to the very concepts and methods allowed (so lacking '*substantial rationality*' *sensu* Karl Mannheim). Mixed farming, the use of local resources, and a focus at a circular economy had been a characteristic of farming for centuries. They now were discarded, not because they had been disproved but because they stood in the way of the projected industrial fertiliser-only agriculture and its scale enlargement. Problems like soil deterioration and eutrophication of surface waters soon started to grow, but with historical agronomy discarded, there was little left to solve them. So it stands to reason that we open up history and look where we can find help. We return again to the mid-nineteenth century.

5. Focus at nutrient solubilisation

As indicated Liebig's influence was not absolute at mid-nineteenth century, not the least because researchers especially in France knew about the quality of De Saussure's experiments and explanations (Petit-Lafitte and de Gasparin were among them). And it was again the common experience with fertile soil from vegetable gardens that induced de Gasparin in 1852 to ask Verdeil and Risler a close investigation of water extracts from a number of such soils, after Liebig had stated that such soil extracts contained hardly any organics at all. Verdeil and Risler [95, 96] refuted that opinion. They showed that the extracts were indeed a source of N-compounds and demonstrated also that these extracts had very considerable mineral solubilising power. Their report was received favourably by the (French) Academy of Sciences [97]. And so the concept of soluble soil organics as central also to mineral nutrient mobilisation for plant uptake—expressed by earlier researchers—was fully corroborated. Risler continued with the research and published his results in the April 1858 issue of the *Archives des sciences de la Bibliothèque universelle* (Genève). He summarised [98] in 1872: **(1)** *this humus not only favours the solubilisation of certain mineral substances that are really needed for plants*, **(2)** *but it also furnishes plants part of their constituent carbon and facilitates the absorption of carbon from the atmosphere.*

The function of humus as a solubiliser and carrier of minerals—phosphate and potassium minerals among them—was implicit in the preparation of bone meal as phosphate fertiliser [99, 100]. Studied early in the century by Lampadius [101–105] and Sprengel, it was further developed by Grandeau ('matière noire') and others. Next in the mid-twentieth century, especially Chaminade [106–112] studied many aspects. Independently it was studied by Åslander, a Swedish researcher who focussed at phosphate fertilisation of the acid peat lands common at Northern latitudes [113–116]. Composted fertiliser with rather small amounts of phosphate sufficed, and no chalk was needed. And it was not fixed by the soil as 'super phosphate' was.

We saw already that the sweetish taste of the soil extract [52, 95–97] was connected with its mineral solubilising power: the solubilisation action of sugar-like compounds was investigated rather early [117]. The second half of the nineteenth century saw steady progress in the characterisation of carbohydrates, etc., but little of it was used by agricultural researchers. If we now jump to the post-Second World War decades, we find growing research in soil carbohydrates and related subjects as well as the changes in soil carbohydrates wrought by straw application but little attention to the interactions with minerals. But in the medical field, biotic (de)mineralisation came in focus early. A distinguished researcher was Carl Neuberg, a prominent biochemist who only recently received the attention he deserved [118]. Early in the twentieth century, he did research on glucuronic acids and related compounds as well as on mineral metabolism, and this starting position brought him ultimately to solubilisation and (de)mineralisation research. Neuberg was a refugee for the Nazi regime who only in 1941, at the age of 61, after great detours managed to reach the USA [119]. He then published a work on mineral solubilisation/(de)mineralisation, with *Remarkable properties of nucleic acids and nucleotides* [120] showing the solubilisation properties of these compounds. [121, 122] focussed at solubilisation and the Ca- and P-cycles in nature but were not noticed by agricultural research. [123, 124]'s work about *Solubilisation of insoluble matter in nature* gave rich (and startling) information but was again not noticed by agricultural research. [125] is a highly useful (last) review, [126] a moving series of lecture demonstrations; all was eminently useful for agricultural research and instruction, but none of it was noticed. *Agricultural research had become isolated*. It is only now catching up: as the reader will know recently humics as carrier of P-compounds became once more the subject of research.

6. Focus on mixotrophy and modelling

De Saussure [22] was well aware of the fact that humus was much too complex to allow component analysis. In 1917 Bottomley [127], by applying the same bicarbonate extractant that De Saussure had used but now without heating, got an extract in which he could show the presence of nucleic acid derivatives (see also [128]). In those decades the group of Schreiner at the USDA Bureau of Soils studied the role in soil-and-plant uptake of organic P-compounds [129, 130]. Schulow [131] and Weissflog and Mengdehl [132] managed to perform such uptake studies under strictly sterile circumstances and found that organic P uptake with maize compared well with uptake of inorganic P. Early on [77] there were already observations of phosphate dissolving power of plant roots; next also phosphate solubilisation by microorganisms was discovered, as was mycorrhizal P-compound uptake (easily disturbed by industrial fertiliser [133, 134]). Altogether we see a P-cycle with biota and organic interactions central and a broad scale of organic P-compounds contributing to plant nutrition. *There are a great number of 'actors' here that in a way are known to the local plant—primarily through its exudation and uptake of organic compounds. But the 'soil P tests' developed in the line of Liebig's 'elements as nutrients' approach offer us no entrance*. To understand that, we take a look at the Olsen available P test.

Olsen [135] has only a few references and misses out even on the most relevant American publications, not only on [136] but also on [137, 138] (refer to [132]). All were clearly expounding plant uptake and assimilation of organic phosphorus compounds. Olsen in fact disqualified his own test, for chemical researchers in the first post-war decades were obliged to consult the German and French literature. Yet, the test has been cited an unbelievable number of times and so helped to shape a virtual world of phosphate fertilisation that ultimately landed us in the extensive eutrophication that we now see everywhere. That Olsen's test is not about real soils is evident from [139–144]. But note that *early on such tests had been found invalid already at the highest soil scientific level* [145–149]. The post-war building was/is without foundations.

As to research on mixotrophy with N-compounds, the first specific investigations were with compounds known also from guano extracts. Wicke and his students perfected methods to study plant uptake of such compounds [150–153]. Uptake and assimilation of several compounds was proved, but with Wicke's successor focussing at animal nutrition, the organic uptake research was not continued. Yet we find uptake studies of organic compounds elsewhere; several reviews were published in the decades around 1900, and up till 1913 a number of PhD theses on the subject were published in Paris (some references in [87]). And at the end of the nineteenth century, the parlance equating 'elements' and 'nutrients' was found inadequate also by researchers who started their career under Liebig's influence. Leading plant physiologist Wilhelm Pfeffer in 1895 reintroduced the chemical concept in his *The election of organic nutrients [by plants]* and stressed that as a rule the plant will take up *organic compounds/complexes* of elements. Pfeffer focussed at lower plants, but his best student Czapek next took a close look at higher plants too. Research intensified, with much work done especially at the USDA Bureau of Soil by Oswald Schreiner's research group (see also [154]).

Czapek was by then the leading authority in the field and gave in 1920 two extensive overviews ([155, 156] fully acknowledging the work at the Bureau of Soils). When Pfeffer died Czapek was appointed in his chair (Leipzig), but Czapek himself died already in 1921, and in a bankrupt Germany there was little prospect for institutional continuity.

The war brought a rupture also in the work at the Bureau of Soils. Lathrop gave an extensive review of research with organic nitrogen compounds in soils and in plant nutrition [157], yet, in 1919, the same journal published Creighton's 'How the nitrogen problem has been solved' [158] that gave extensive accounts of nitrogen fixation industries without a word on soil-and-plant research. Moreover the nitrogen fixation unit that did research in nitrogen fixation for explosives production was positioned in the Bureau of Soils and dwarfed Schreiner's group (that now was obliged to focus far more on industrial fertilisers). Yet, Schreiner already in 1912 introduced an extended N-cycle [159, 160] pointing to thermodynamic a.o. advantages for the plant if it absorbs organic N breakdown products of soil organics before their 'mineralisation' (see also [161]). In the present this is a lively research subject, after researchers found organic N to contribute greatly to plant nutrition in boreal and arctic regions. In the soil there is a dazzling number of 'actors'—think also of biological nitrogen fixation—and of organic compounds that contribute to the soil-and-plant N-cycle. *Again it is the local plant that 'knows' about it all—and future models will have to start from that fact.* But neither the government nor the industry is at home in this soil world, and their orders—and fertiliser-centred models—are without power there.

7. Summary and outlook

There are many more important subjects waiting for exposition, from the high-level field experiments of Dehérain with their renewed attention to soils and organics and to the important and diverse roles of biochars in 'traditional' European agricultures. But the subjects touched upon will suffice to show that our post-war 'modern agriculture' was not for real. Its agronomy denied mixotrophy, organic solubilisation, and other organic-mineral interactions that, yet, had their roots in soil, farmer practices, and peer-reviewed science. Realising that much we are free again to effect a change to an agronomy for sustainable agricultures that starts from respecting both the soil and the work of those who interact with it daily. The suggestion that orders from the industry and government can make crops grow was at the centre of post-war policies. Its agronomy was ready-made for the purpose—think of its soil nutrient tests—but plants, earthworms, and microorganisms did not listen, and we ended up with global eutrophication and soil deterioration. Redevelopment is urgent and possible—not easy—and we are fortunate that there is a treasure of historical knowledge and experience that can assist us.

Author details

Jozef Visser
Retired, Utrecht, The Netherlands

*Address all correspondence to: joost.visser@ziggo.nl

References

[1] Carrière J. Ihre Briefe von 1831-1845 mit erläuternden Einschaltungen aus. In: Briefe von Liebig und Wöhler. München/Leipzig: Berzelius und Liebig; 1893

[2] Krzymowski R. Philosophie der Landwirtschaftslehre. Stuttgart; 1919

[3] Soubeiran E. Analyse chimique de l'humus et rôle des engrais dans l'alimentation des plantes. Journal de Pharmacie et de Chimie,Série 3 Tome. 1850;**17**:321-334

[4] Soubeiran E. Über den Humus und die Wirkungen des Düngers bei der Ernährung der Pflanzen. Journal für praktische Chemie. 1850;**50**:291-305

[5] Soubeiran E. Analyse chimique de l'humus et rôle des engrais dans l'alimentation des plantes II Analyse de quelques engrais. Journal de Pharmacie et de Chimie, Série 3 Tome. 1850;**18**:5-20

[6] Malaguti MJ. Note sur l'absorption des ulmates solubles par les plantes. Annales de Physique et de Chimie. 1852;**3**(34):140-143; also in Comptes Rendus de l'Académie des Sciences de Paris T.34, 112-114

[7] Corenwinder M. Études sur les fonctions des racines des végétaux. Annales de sciences naturelles, Série 5 T. 1868;**9**:63-69

[8] Wolff E. Die naturgesetzliche Grundlagen des Ackerbaues. Zweite Auflage, Leipzig; 1854

[9] Johnson J. Besprechung von E. Wolff's 'Die Naturgesetzlichen Grundlagen des Ackerbaues'. In: Mittheilungen der K. 2te Auflage ed. Vol. Band 1 and 2, 1854. St. Petersburg: Freien Ökonomischen Gesellschaft; 1855. pp. 234-251

[10] Johnson J. Über die Werthbestimmung des Stalldüngers. In: Mittheilungen der K. St. Petersburg: Freien Ökonomische Gesellschaft; 1857. pp. 329-339

[11] Risler E. Sur l'enseignement agricole en Allemagne. Journal D'agriculture Pratique Année 26. 1862;**1**:365-369

[12] Risler E. Histoire des théories agronomiques. Journal D'agriculture Pratique Année 26. 1862;**2**:348-351

[13] Stohmann. Liebig's Beziehungen zur Landwirthschaft. Journal für praktische Chemie. 1873-1874:458-476

[14] Schmalz F. Erfahrungen im Gebiete der Landwirthschaft. Bd. 1-6 ed. Leipzig; 1814-1824

[15] Schmalz F. Offenes Sendschreiben: An den Herrn Professor und Ritter Dr. Julius Liebig. Allgemeine Landwirtschaftliche Monatsschrift. 1841;**4**:44-60

[16] Liebig J. Abfertigung des Herren Dr. Gruber in Wien und Dr. Sprengel, in Beziehung auf ihre Kritiken meines Werkes: "die organische Chemie". Annalen der Chemie und Pharmacie. 1841;**38**:216-256

[17] Sprengel C. Auszüge aus Prof. Liebigs "organische Chemie in ihrer Anwendung auf Agricultur und Physiologie", nebst Bemerkungen vom Redacteur. Allgemeine Landwirthschaftliche Monatsschrift. 1840;**2**:171-194

[18] Liebig J. Die organische Chemie in ihrer Anwendung auf Agricultur und Physiologie. Braunschweig; 1840

[19] Sprengel C. Über Pflanzenhumus, Humussäure und humussauren Salze. Archiv für die gesammte Naturlehre. 1826;**8**:145-220

[20] Schmalz F. Aphorismen aus der Pflanzenernährungskunde. Allgemeine

landwirtschaftliche Monatsschrift. 1842;**7**:60-84

[21] Liebig J. Bemerkungen zur vorstehende Versuchen de Saussure's. Annalen der Chemie und Pharmazie. 1842;**42**:291-298

[22] De Saussure T. Sur la nutrition des végétaux. Bibliothèque Universelle de Genève, Nouvelle Série. 1841;**36**:340-355

[23] De Saussure T. Über die Ernährung der Pflanzen. Annalen der Chemie und Pharmazie. 1842;**42**:275-291

[24] Trinchinetti A. Sulla facolta assorbente delle radici de'vegetabili. Milano; 1843. pp. 81

[25] Johnson J. Von der Nahrung der Culturpflanzen, Mittheilungen der K. St. Peterburg: Freien Ökonomischen Gesellschaft; 1844. pp. 185-210

[26] Schlechtendal. Von der Nahrung der Culturpflanzen. In: Inaugural-Dissertation von Jacob Johnson, St. Petersburg, 1844, 8 & 56 S Botanische Zeitung 3. 1845. pp. 632-634

[27] Heuchert B, Braun U, Tkach N, Marx D, Röser M. Biography of D.F.L. von Schlechtendal and type material of his new taxa preserved in the herbarium of Martin Luther University Halle-Wittenberg (HAL) and other botanical collections. In: Schlechtendalia. Vol. 31. 2017. p. 143

[28] von Mohl H. Dr. Justus Liebig's Verhältnis zur Pflanzenphysiologie. Tübingen; 1843

[29] von Mohl H. Die Chemie in ihrer Anwendung auf Agricultur und Physiologie. Von Julius Liebig. 5te umgearb. u. sehr vermehrte Aufl. 1843. Botanische Zeitung. 1844;**3**:358-361

[30] Fürnrohr AE. Dr. Justus Liebig's Verhältnisse zur Pflanzenphysiologie. In: von Mohl H, editor. Flora. Jahrgang 26 Bd ed. Vol. 1. 1843. pp. 67-68

[31] von Mohl H. Grundzüge der Anatomie und Physiologie der vegetabilischen Zelle. Braunschweig; 1851

[32] Volhard J. Justus von Liebig. Band 1 & 2 ed. Leipzig; 1909

[33] Le Petit-Lafitte A. De la connaissance des terres cultivées. Bordeaux; 1845

[34] Le Petit-Lafitte A. Étude des terres arables. Paris; 1862

[35] Schubert F. Handbuch der Forstchemie. Leipzig; 1848

[36] Fresenius R. Lehrbuch der Chemie für Landwirthe. Braunschweig: Forstmänner und Cameralisten; 1847

[37] Strecker A. Untersuchungen über die chemische Constitution der Hauptbestandtheile der Ochsengalle. Der philosophische Fakultät der Ludwigsuniversität Giessen; 1848. Vorgelegt zur Erlangung der venia legendi

[38] Strecker A. Thesen Behufs der Erlangung der venia legendi. Ludwigsuniversität Giessen. öffentlich [zu] vertheidigen; 1848

[39] Landes-Oeconomie-Collegium. Bericht über die auf Veranlassung des Königlichen Landes-Oeconomie-Collegium ausgeführten Versuche mit Liebig's künstlichem Dünger. Annalen der Landwirthschaft in den Königlichen Preußischen Staaten. 1847;**10**:371-392

[40] Magnus E. Versuche, betreffend die Erschöpfung des Bodens, welche das Königliche Landes-Oeconomie-Collegium veranlasst hat. Annalen der Landwirthschaft in den Königlich Preußischen Staaten. 1849;**14**:179-216

[41] Henrici FG. Bemerkungen über die neuen die Landwirthschaft betreffenden chemischen Briefe des Herrn von Liebig. Göttingen; 1858

[42] Jacobi V. Freiherr von Liebig als unberechtigt zu entscheidendem Urteil über Praxis und Unterrichtswesen in der Landwirthschaft. Land- und forstwirthschaftlich, didaktisch, nationalökonomisch und historisch nachgewiesen. Leipzig; 1862

[43] Koppe JG. Mittheilungen über die Geschichte des Ackerbaues in Nord-Deutschland, besonders in Preußen, im achtzehnten und neunzehnten Jahrhundert, zur Prüfung der Frage: ob Gründe vorliegen, der neueren Landwirthschaft Schuld zu geben, daß sie ein Raub-System befolge? Berlin; 1860

[44] Arnd K. Justus Liebig's Agrikulturchemie und sein Gespenst der Bodenerschöpfung. Ein Beitrag zur Prüfung seiner Verurtheiliung der heutigen europäischen Landwirthschaft. Frankfurt a. Main; 1864

[45] Fraas G. Bavaria rediviva! Ein Beitrag zur Lehre vom Völkeruntergang durch Bodenerschöpfung. München; 1865

[46] Au J. J.v.Liebig's Lehre von der Bodenerschöpfung und die nationalökonomischen Bevölkerungstheorien dargelegt und kritisch untersucht. In: Heidelberg. 1869

[47] Waltz G. Mitteilungen aus Hohenheim. 1865;**6**

[48] Mulder GJ. Die Chemie der Ackerkrume 1861-1863;1-3. Übersetzung und Erläuterungen Johannes Müller, E. Gross, Berlin

[49] van Bemmelen JM. Bijdrage tot de wetenschappelijke biographie van G.J. Mulder. Historisch-kritische beschouwing van zijn werk: De Scheikunde der bouwbare Aarde, Verhandelingen Koninklijke Akademie van Wetenschappen te Amsterdam (Eerste Sectie). 1901;**VII**(7)

[50] Fraas G. Buch der Natur für Landwirthe oder landwirthschaftliche Naturkunde. München; 1860

[51] Fraas G. Das Wurzelleben der Kulturpflanzen und die Ertragssteigerung. Nebst einer Tabelle der Bewurzelungssysteme der Kulturpflanzen. Leipzig; 1870

[52] Estienne C, Liebault J, Sebisch M. Fünftzehn Bücher vom Feldbaw. Straßburg; 1592

[53] Gallo A. Le vinti giornate dell'agricoltura. Venetia; 1628

[54] Coler J. Oeconomia ruralis et domestica I, II. Maynz; 1645

[55] Anonymous. Der Sorgfältige Haushalter. Osnabrück; 1674

[56] Beste A. Qualitative Bodenbeurteilung zur Erhaltung der Bodenfunktionen für Produktion, Gewässerschutz und Hochwasservermeidung. Buro für Bodenschutz und ökologische Agrarkultur; 2005

[57] Beste A. Landwirtschaftlicher Bodenschutz in der Praxis. Grundlagen, Analyse, Management. Berlin; 2005

[58] McGarry D. Methodology of a Visual Soil–Field Assessment Tool. FAO; 2005

[59] Emmett-Booth JP, Forristal PD, Fenton O, Ball BC, Holden NM. A review of visual soil evaluation techniques for soil structure. Soil Use and Management. 2016;**32**:623-634

[60] Leopizzi S, Gondret K, Boivin P. Spatial variability and sampling requirements of the visual evaluation of soil structure in cropped fields. Geoderma. 2018;**314**:58-62

[61] Heresbach/Googe. The Whole Art and Trade of Husbandry. London; 1614

[62] Rhagor D. Pflanz-Garten. Bern; 1650

[63] Von Roll T. Neues Blumen-Büchlein. Constanz; 1669

[64] Stromer von Reichenbach WA. Die edle Garten-Wissenschaft. Nürnberg; 1671

[65] Böckler GA. Nützliche Hauß- und Feld-Schule. Nürnberg; 1678

[66] Winiwarter V. Soils and Societies: Perspectives from Environmental History. Isle of Harris: White Horse Press; 2010

[67] Winiwarter V. Environmental history of soils. In: Agnoletti et al., editors. World Environmental History. 2011

[68] Lal R. Restoring soil quality to mitigate soil degradation. Sustainability. 2015;7:5875-5895

[69] Mayer JF. Abhandlung über die Aufgabe einer faßlichen Bescheibung der verschiedenen Erdarten etc und einer Anzeige wie selbige durch ihre Vermischung fruchtbar zu machen? In: Abhandlungen und Beobachtungen durch die Ökonomische Gesellschaft zu Bern gesammelt, Abhandlung IV. Vol. 9(1). 1768

[70] von Hartmann L. Abhandlung von der Erkenntniß und Verbesserung der Erde. Burghausen; 1772

[71] Lüders PE. Drey Bedenken über die Erforschung, die Beurtheilung, und den Gebrauch der Erd-Arten. Flensburg; 1764

[72] Sprenger B. Kurzer doch gründlicher Begriff des gesamten Feldbaues. Stuttgart; 1764

[73] Andreae. Abhandlung über ein beträchtlicher Anzahl Erdarten und von derselben Gebrauch für den Landwirth. Hannover; 1769

[74] Petersen. Kurzgefaste Abhandlungen über verschiedene den Ackerbau betreffende Gegenstände. Flensburg; 1769

[75] Home F. Von den Grunsätzen des Feldbaues und des Wachsthumes der Pflanzen. Edinburgh; English original

[76] Marci A. Mémoire sur l'utilité des engrais artificiels. Mémoires de l'Aacadémie Impériale et Royale des Sciences de Bruxelles T. 1777;**3**:47-74

[77] Ziel L. Die landwirtschaftliche vegetabilische Produktionslehre nach den Grundsätze der Naturwissenschaft und Erfahrung. München; 1830

[78] Schwerz JN. Anleitung zum praktischen Ackerbau. Vol. 1. Stuttgart & Tübingen; 1823

[79] Sprengel C. Die Bodenkunde oder die Lehre vom Boden. Leipzig; 1837

[80] Kneuttinger. Die Gahre oder Gährung des Ackers. Zeitschrift des landwirtschaftlichen Vereins in Bayern. 1863-1865;**53**:534-544. 54, 347-352; 55, 99-117

[81] Görbing J. Die Grundlagen der Gare im praktischen Ackerbau. Vol. 1 & 2. Hannover: Landbuch-Verlag; 1947

[82] Dittrich H. Untersuchungen über die Bodengare. Bodenkunde und Pflanzenernährung. 1940;**16**:16-50

[83] Sekera F. Der allgemeine Bauplan der Bodenstruktur und die Dynamik der Bodengare. Zeitschrift für Pflanzenernährung, Düngung, Bodenkunde. 1951;**52**:57-60

[84] Sekera F. Gesunder und kranker Boden. Ein praktischer Wegweiser zur Gesundhaltung des Ackers (Kap. 1 Bodengare), Dritte, völlig neubearb. Berlin: Auflage; 1951

[85] Fruwirth C. Die Ackerbaulehre. Berlin: Paul Parey; 1921

[86] Sauerlandt W. Grundlagen der Bodenfruchtbarkeit. In: Humusdüngung und Bodengare. Lüneburg: Metta Kinau Verlag Nachf.; 1948

[87] Paungfoo-Lonhienne C, Visser J, Lonhienne TGA, Schmidt S. Marschner review: Past, present and future of organic nutrients. Plant Soil. 2012;**359**:1-18

[88] Visser J. Down to Earth [PhD thesis]. University Wageningen; 2010

[89] Johnson J. Über Gemengsaaten von Erbsen und Hafer, Mittheilungen der K. Freien Ökonomische Gesellschaft; 1853. pp. 50-51

[90] Stoddard FL, Hovinn S, Kontturi M, Lindtstrom K, Nikånen A. Legumes in Finnish agriculture: History, present status and future prospects. Agricultural and Food Science. 2009;**18**:191-2005

[91] Sohn López-Forment IM. Ecological and socio-cultural dynamics of traditional and legume-based Milpa agriculture in S.E. Mexico [PhD]. Ohio State University; 2000

[92] Benítez M, Fornoni J, García-Barrios L, López R. Dynamical networks in agroecology: The milpa as a model system. In: Benítez et al., editors. Frontiers in Ecology, Evolution, and Complexity. 2014

[93] Melsted SW. New concepts of management for corn belt soils. Advances in Agronomy. 1954;**6**:121-143

[94] Joffe JS. Green manuring viewed by a pedologist. Advances in Agronomy. 1955;7:142-188

[95] Verdeil F, Risler E. Recherches sur la composition des matières solubles extraites, par l'eau, des terres fertiles. Comptes rendus de l'Académie des Sciences. 1852;**35**:95-99

[96] Verdeil F, Risler E. Matières solubles des terres fertiles. Journal D'agriculture Pratique. 1852;**3**(5):340-342

[97] Boussingault P, Gasparin D. Rapport sur un Mémoire de MM. Verdeil et Rissler, intitulé: Recherches sur la composition des matières solubles eaxtraites par l'eau des terres fertiles. Comptes rendus de l'Académie des sciences. 1853;**36**:765-768

[98] Risler E. Quotes in: A. de Céris, Chronique agricole. Journal D'agriculture Pratique. 1872;**36**(1):378-379

[99] Cohn W. Über das Knochenmehl und seine Anwendung als Düngemittel in chemischer und landwirtschaftlicher Beziehung; 1858

[100] Cohn W. Über die Zubereitung des Knochenmehl zur Düngung, Mittheilungen der K. St. Petersburg: Freien Ökonomische Gesellschaft; 1862. pp. 218-222

[101] Lampadius WA. Über das Knochenmehl, als Düngungsmittel für Gärten und Aecker. Journal für Technische und Ökonomische Chemie. 1828;**1**:23-28

[102] Lampadius WA. Über die Wirkung durchglüheter erdiger Massen als Beförderungsmittel der Vegetation, mit besonderer Berücksichtigung der hieher gehörigen Erfahrungen des Ritter von Schindler. Journal für Technische und Ökonomische Chemie. 1833;**18**:322-331

[103] Lampadius WA. Agronomische Versuchen mit künstlichen Düngemittel angestellt im Jahre 1835. I. Fortgesetzte Versuche und Erfahrungen über die Wirkung humussauerer Basen als

Düngungsmittel. Journal für Praktische Chemie. 1835;**5**:433-450

[104] Lampadius WA. Agronomisch-chemische Versuche und Erfahrungen. Journal für praktische Chemie. 1838;**15**:338-349

[105] Lampadius WA. Fortsetzung der Mittheilung bestätigender Erfahrungen über die Wirkung humussauren Basen, vorzüglich der aus Torf bereitete, als Düngmittel. Journal für praktische Chemie. 1840;**20**:267-271

[106] Chaminade R. Sur l'existence dans les sols de complexes phospho-humiques. Comptes Rendus Académie des Sciences (Paris). 1943:235-237

[107] Chaminade R. Sur les propriétés des complexes phospho-humiques des sols. Comptes Rendus Académie des Sciences (Paris). 1943:275-278

[108] Chaminade R. Les formes du phosphore dans le sol. Nature et rôle des complexes phospho-humiques. Paris: Dunod; 1944

[109] Chaminade R. Sur l'existence et les conditions de formation des composés d'adsorption phospho-humiques. Comptes Rendus Académie des Sciences (Paris). 1946:168-170

[110] Chaminade R, Blanchet R. Action stimulante de l'humus sur le développement et la nutrition minérale des végétaux dans le sol. Comptes Rendus Académie des Sciences (Paris). 1953:119-121

[111] Chaminade R. Influence de la matiere organique humifiée sur l'efficacité de l'azote. Zeitschrift für Pflanzenernährung, Düngung, Bodenkunde. 1959;**84**:22-25

[112] Chaminade R. Rôle specifique de la matière organique sur la nutrition et le rendement des végétaux. In: Pontificia Academia Scientiarum. Vol. 1968. 1968. pp. 777-800

[113] Åslander A. The availability of phosphates after standard fertilization. In: Transactions of the 4th International Congress of Soil Science (Amsterdam), Volume II. 1950. pp. 158-163

[114] Åslander A. Standard fertilization and liming as factors in maintaining soil productivity. Soil Science. 1952;**74**:181-195

[115] Åslander A. Standard fertilization and the quality of crops. Soil Science. 1952;**74**:431-442

[116] Åslander A, Armolik N. The influence of organic materials on the potassium fixation in the soil. Transactions of the Royal Institute of Technology, Stockholm. 1964;**236**

[117] Peligot E. Recherches sur la nature et les propriétés chimiques des sucres. Annales de Chimie et de Physique. 1838;**67**:113-177

[118] Conrads H, Lohff B. Carl Neuberg, Biochemie, Politik und Geschichte. Lebenswege und Werk eines fast verdrängten Forschers. Ferdinand Steiner Verlag; 2006

[119] Nordwig A. Carl Neuberg: Fate of a Jewish biochemist in the Third Reich. TIBS. 1984:498-499

[120] Neuberg C, Roberts IS. Remarkable properties of nucleic acids and nucleotides. Archives of Biochemistry. 1949;**20**:185-210

[121] Neuberg C, Grauer A. The problem of the solubilization and precipitation and the calcium and phosphorus cycles in cavern formation. Experientia. 1957;**10**:391-393

[122] Neuberg C, Grauer A, Kreidl M, Lowy H. The role of the carbamate

reaction in the calcium and phosphorus cycles in nature. Archives of Biochemistry and Biophysics. 1957;**70**:70-79

[123] Mandl I, Grauer A, Neuberg C. Solubilization of insoluble matter in nature. I. The part played by salts of adenosinetriphosphate. Biochimica et Biophysica Acta. 1952;**8**:654-663

[124] Mandl I, Grauer A, Neuberg C. Solubilization of insoluble matter in nature. II. Part played by salts of organic and inorganic acids occurring in nature. Biochimica et Biophysica Acta. 1953;**10**:540-569

[125] Mandl I, Neuberg C. Solubilization, migration, and utilization of insoluble matter. In: Nord FF, editor. Advances in Enzymology and Related Areas of Molecular Biology. Vol. 17. 1956. pp. 135-159

[126] Pro memoria Carl Neuberg, Lecture-Demonstration by Carl Neuberg. Experientia. 1957;**13**:451-453

[127] Bottomley WH. The isolation from peat of certain nucleic acid derivatives. Proceedings of the Royal Society B. 1917;**90**:39-44

[128] MacLean AA. Extraction of organic phosphorus from soils with bicarbonate. Canadian Journal of Soil Science. 1965;**45**:165-170

[129] Auten JT. Organic phosphorus of soils. Soil Science. 1923;**16**:281-294

[130] Schreiner O. Organic phosphorus in soils. Agronomy Journal. 1923;**15**:117-124

[131] Schulow IW. Versuche mit sterilen Kulturen höherer Pflanzen. Berichte der deutschen botanischen Gesellschaft. 1913;**31**:97-121

[132] Weissflog J, Mengdehl H. Studien zum Phosphorstoffwechsel in der höhern Pflanze. III. Aufnahme und Verwertbarkeit organischer Phosphorsäureverbindungen durch die Pflanze. Planta: Archiv für wissenschaftliche Botanik. 1933;**19**:182-241

[133] Breuillin F, Schramm J, Hajirezaei M, Ahkami A, Favre P, Druege U, et al. Phosphate systemically inhibits development of arbuscular mycorrhiza in *Petunia hybrida* and represses genes involved in mycorrhizal functioning. The Plant Journal. 2010;**64**:1002-1017

[134] Aguacil M d M, Lozano Z, Campoy MG, Roldán A. Phosphorus fertilisation management modifies the biodiversity of AM fungi in a tropical savanna forage system. Soil Biology and Biochemistry. 2010;**42**:1114-1122

[135] Sterling HO, Cole CV, Watanabe FS, Dean LA. Estimation of available phosphorus in soils by extraction with sodium bicarbonate. In: United States Department of Agriculture Circular No. 939. 1954

[136] Heck AF. Studies on the assimilation of organic phosphorus by plants [thesis]. University of Illinois; 1916

[137] Rodney Bertramson B. Organic forms of phosphorus in the phosphorus nutrition of plants [thesis]. Oregon State College; 1941

[138] Bower CA. Studies on the forms and availability of soil organic phosphorus [dissertation]. Iowa State University; 1949

[139] Steffens D, Leppin T, Luschin-Ebengreuth N, Yang ZM, Schubert S. Organic phosphorus considerably contributes to plant nutrition but is neglected by routine soil-testing methods. Journal of Plant Nutrition and Soil Science. 2010;**173**:765-771

[140] Wening A. Pflanzenverfügbares organisch gebundenes Phosphat in Abhängigkeit von Pflanzenart und Bodeneigenschaften [dissertation Gießen]; 2016

[141] Trapp S, Eggen T. Simulation of the plant uptake of organophosphates and other emerging pollutants for greenhouse experiments and field conditions. Environmental Science and Pollution Research. 2012;**20**:4018-4029

[142] Paraskova JV. Organic phosphorus speciation in environmental samples. Method development and applications [dissertation]. Uppsala University; 2014

[143] Rubæk GH. Validity and analytical robustness of the Olsen soil P test and other agronomic soil P tests used in Northern Europe. In: Report Nr. 071. DCA Danish Centre for Food and Agriculture, Aarhus University; 2015

[144] Rubaek GH, Kristensen K. Protocol for bicarbonate extraction of inorganic phosphate from agricultural soils. In: Report nr. 102. DCA Danish Centre for Food and Agriculture, Aarhus University; 2017

[145] Internationale Gesellschaft für Bodenkunde. Erster Bericht über die Arbeiten und über die Tagung der Arbeitsgemeinschaft zur Prüfung der Laboratoriumsmethoden für die Bestimmung des Kali-und des Phosphorsäurebedürfnisses der Böden. In: Mistcherlich VEA, editor. Königsberg (Pr.) 12. bis 19. Juli 1936, Herausgegeben; Königsberg (Pr.). 1937

[146] Internationale Gesellschaft für Bodenkunde. Zweiter Bericht. Verhandelt in Stockholm am 5. Juli 1939, Herausgegeben von Eilhard Alfred Mistcherlich, Königsberg (Pr.); 1939

[147] Eilh. Afred Mitscherlich. Physiologische Bodenkunde; 1948

[148] Mitscherlich EA. Die “Chemische Bodenanalyse”. Zeitschrift für Pflanzenernährung, Düngung, Bodenkunde. 1948;**87**:97-103

[149] Mitscherlich EA. Eine Ergänzung zum zweiten Bericht der Arbeits-gemeinschaft zur Prüfung der Laboratoriumsmethoden für die Bestimmung des Kali- und Phosphorsäurebedürfnisses der Böden. Bulletin of the International Society of Soil Science. 1954;**5**(1954):10-19

[150] Hampe W. Über die Ammoniaksalze, Harnsäure, Hippursäure und Glycin als stickstoffhaltige Nahrungsmittel der Pflanzen, Nachrichten von der Königl. Gesellschaft der Wissenschaften und der Georg-August-Universität zu Göttingen. 1868;**3**:94-105

[151] Wicke W. Der Harnstoff als stickstoffhaltiges Pflanzen-Nahrungsmittel. Nachrichten Königliche Gesellschaft der Wissenschaften und der Georg-Augusts-Universität Göttingen. 1865:352-356

[152] Wicke W. (mit Wagner, Mölln) Vegetations-Versuche mit phosphorsaurem Ammon, Hippursäure, Glycin und Kreatin, Nachrichten von der Königl. Vol. 4. Gesellschaft der Wissenschaften und der G.A. Universität zu Göttingen; 1869. pp. 43-57

[153] Wagner P. Vegetations-Versuche über die Stickstoff-Ernährung der Pflanzen [Inaugural-Dissertation]. Universität Göttingen; 1869

[154] Brigham RO. Assimilation of Organic Nitrogen by *Zea mays* and the influence of *Bacillus subtilus* on such assimilation [thesis]. University of Michigan; 1917

[155] Czapek F. Die organische Ernährung bei höheren grünen

Pflanzen. Die Naturwissenschaften. 1920;**12**:226-232

[156] Czapek F. Resorption organischer Stickstoffverbindungen durch Phanerogamenwurzeln. In: Friedrich Czapek 1920 Biochemie der Pflanzen. 2e Auflage ed. Vol. 2. 1920. pp. 318-321

[157] Lathrop C. The organic nitrogen compounds of soils and fertilizers. Journal of the Franklin Institute. 1917;**183**:169-206; 303-321; 465-498

[158] Jermain H, Creighton M. How the nitrogen problem has been solved. Journal of the Franklin Institute. 1919;**187**:377-408; 599-610; 705-735

[159] Schreiner O, Skinner JJ. Nitrogenous soil constituents and their bearing on soil fertility. USDA Bureau of Soil Bulletin. 1912;**87**

[160] Schreiner O. Organic soil constituents in their relation to soil fertility. In: Eighth International Congress of Applied Chemistry, Section VII: Agricultural Chemistry. 1912. pp. 231-245

[161] Schreiner O, Skinner JJ. Experimental study of the effect of some nitrogenous soil constituents on growth. Nucleic acid and its decomposition products. The Plant World. 1913;**16**:45-60

[162] Karstens B. Pluralism within parameters: Towards a mature evaluative historiography of science [PhD thesis]. Leiden University; 2015

[163] Bouterse J. Nature and history: Towards a hermeneutic philosophy of science [PhD thesis]. Leiden University; 2016

[164] Uekötter F. Did they know what they were doing? GHI Bulletin Supplement. 2006;**2006**:145-166

[165] Uekötter F. Perspektiven einer Weltgeschichte des landwirtschaftlich genutzten Bodens. In: Herrmann B, editor. Beiträge zum Göttinger Umwelthist, Kolloquium 2004-2006. 2007. pp. 29-40

[166] Uekötter F. Virtuelle Böden. Über Konstruktion und Destruktion des landwirtschaftlichen Bodens in den Agrarwissenschaften. Zeitschrift für Agrargeschichte und Agrarsoziologie. 2007;**55**:23-42

[167] Uekötter F. Im Schatten von Liebig. Das Wissen um den Boden, eine Verlustgeschichte. In: Der kritische Agrarbericht. 2010. pp. 261-265

[168] Uekötter F. The magic of one. Reflections on the pathologies of monoculture. RCC Perspectives. 2011, 2011;**2**:20

[169] Uekötter F. Europäische Geschichte Online (ERO). Boden; 2012

Chapter 2

The State of the Soil Organic Matter and Nutrients in the Long-Term Field Experiments with Application of Organic and Mineral Fertilizers in Different Soil-Climate Conditions in the View of Expecting Climate Change

Ladislav Menšík, Lukáš Hlisnikovský and Eva Kunzová

Abstract

Soil organic matter (SOM) plays an important role in the terrestrial ecosystems and agroecosystems. Changes in the agricultural sector in the countries of the Central and Eastern Europe (the Czech Republic, Slovakia, Poland, etc.) within the past 25 years have negatively affected the SOM and contributed to the soil degradation. The aim of this chapter is the evaluation of the long-term application of mineral fertilizers and farmyard manure: the Control (without fertilization), farmyard manure (FYM + 0), FYM accompanied with NPK (FYM + N_3PK), and FYM with mineral nitrogen FYM + N (FYM + N_2), on the essential chemical properties of the soil and yield of the fundamental arable crops in the long-term field experiments, established in different soil and climate conditions (black soils, brown soils, cambisols, altitude ranging from 260 to 650 m a.s.l.) of the Czech Republic in 1955, using the modern multi-criteria statistical methods (PCA, FA, CLU, etc.). The long-term and regular application of organic manure and organic manure with mineral fertilizers (FYM + N_3PK and FYM + N_2) optimize the soil characteristics, stabilize crop and feedstuff production, and increase the adaptation potential of the soil in the Czech Republic, which is supposed to be weakened due to the expected changes of the environmental conditions in the near future.

Keywords: soil organic matter, nutrients, long-term field experiments, different soil-climate conditions, multi-criterial evaluation, PCA, FA, CLU

1. Introduction

In the current world, facing the climate and demographic changes, agriculture plays an important role not only as the food producer, feeding the rapidly increasing world population, but also as a feedstuff producer and also an important factor affecting the world climate [1, 2]. The worldwide agricultural production has increased significantly over the last 50 years, but the future demand for cereals, feedstuff, and renewable energy sources will increase considerably as a result of the growing population [3–5]. On the other hand, however, the agricultural crop yields have declined globally over the past 20–30 years [6] due to global warming, and the results of the model studies suggest that climate change will further reduce the yield potential of food and feedstuff, including maize [7–12]. It is supposed that Europe is not going to be affected to such extent as other parts of the world [13] and the impact of climate change in Europe will not affect EU countries equally. It is assumed that the most affected crop in Europe will be maize [12].

Soil organic matter (SOM) plays an important role in terrestrial ecosystems and agroecosystems [14]. It is an important factor related to the three components of soil quality and fertility [15]. From the chemical point of view, SOM largely determines, together with clay minerals, the cation exchange (and anion retention) capacity of soil, pH buffering capacity, and the retention of inorganic and organic pollutants or toxic elements [16, 17]. From the physical point of view, SOM is crucial in determining the soil structure and thereby ultimately controlling soil erosion, water infiltration and holding capacity, and habitat provision for plant roots and soil organisms [18]. From the biological point of view, SOM is a primary source of energy for soil microorganisms and thus the whole soil food net, as well as a source of major nutrients, most notably nitrogen, phosphorus, and sulfur, for plants and the soil biota.

Current status and changes in soil organic carbon stock, in response to agronomic and climatic conditions, become extremely important today [19]. There is an effective strategy to mitigate global climate change by increasing carbon stock in soil [20, 21]. The level and balance of soil organic carbon and mineral nutrients are also the main criterion of agricultural sustainability [22, 23]. Sustainability depends on soil ability to maintain productive and other nonproductive functions (biodiversity, hygienic, environmental, etc.). Sustainable soil management systems require the proper choice of crop rotation system, agricultural practices, carbon stock, as well as a supply of nutrients to reach higher productivity [24].

The intensive use of soil is essential, but it must be associated with conservation practices [25]. One of the main consequences of agricultural land degradation is the C depletion in soils [26]. Loss of soil carbon degrades these services, decreasing crop yields and environmental and market value of the soil. Land management can also enhance soil carbon content by optimal crop rotation and fallow cover crops, organic matter application, optimized fertilization application, and tillage systems [15].

Deterioration of natural sources quality is leading to a negative influence on soil quality (degradation), which agricultural activity depends on. This is also evident in the Czech Republic during the past 25 years [27]. Soil quality deterioration is primarily caused by four major factors: (a) Changes in the structure of cultivated crops and crop rotations (reduction of the share of perennial fodder crops (alfalfa, clover [index 1990/2015: 35%]) and cereals [index 1990/2015: 84%] on behalf of market crops/rapeseed [index 1990/2015: 343%]). (b) Significant reduction of animal husbandry (cattle [index 1990/2015: 40%], pigs [index 1990/2015: 31%], and sheep [index 1990/2015: 50%]) with large differences between regions (large areas without animal husbandry). The average charge of the agricultural land is currently 0.37 LUs per 1 ha^{-1} [28]. (c) Low inputs of organic manures (farmyard manure, slurries, etc.) and, as a consequence, low inputs of organic matter to the soil (the average N, P, and K intake in manures [index 1990/2015: 50, 50, and 50%, respectively]). (d) Reduced application of mineral fertilizers (P, K) [index 1990/2015: 17; 15%] and increased use of mineral nitrogen [index 1990/2015: 101%], resulting in higher soil acidification [28]. Zhang et al. [29] and Ren et al. [30] quoted that decreasing application of manures and organic fertilizers influenced not only stable organic compounds but also soil microorganisms and nutrients regimes. From the soil quality point of view, organic matter (farmyard manure, slurries, and high-quality compost) play an irreplaceable role during the humification process, during the formation of stable humus fractions, and in the fertilization management [27, 31]. Thus, continuous application of balanced fertilizers is necessary for sustaining soil fertility and productivity of crops [32].

The aim of this study was to estimate the effect of long-term application of different organic manures and mineral fertilizers (Control, FYM + 0, FYM + N_3PK, and FYM + N_2) on soil properties in different soil-climate conditions in the Czech Republic. Three long-term field experiments were established in 1956. Basic soil reaction, carbon and nitrogen content, and available nutrient content were analyzed during 2012–2015 and evaluated using the modern multi-criteria statistical methods (PCA, FA, etc.).

2. Organic and mineral fertilizers

Organic fertilizers represent a wide group of materials derived from agricultural by-products, plants, and animal husbandry, such as manures and litters. Organic fertilizers are essential especially for soil microorganisms, which decompose fertilizer's matter to grow and release fertilizer's nutrients into the soil environment. The nutrients then can be utilized by arable crops and support theirs grow and development.

The first systematic use of organic fertilizers is connected with the Neolithic period in the area of the Fertile Crescent approximately 12, 500 years ago. During the Neolithic Revolution, human population and society started changing their habits from essentially hunters and gatherers to farmers and breeders. Together with the establishment of the first crop production, they also started with the domestication of goats, sheep, and later cattle. Waste pits became commonly used in the settlements approximately 6000 years ago to store biologically degradable wastes, stored for the use in the agriculture. So began the use of organic fertilizers by the human population, and to this day, the sense and principle have not changed. Organic manures and fertilizers have played and play several important roles in today's agriculture. They serve as a feedstuff for soil microorganism and significantly affect soil's biodiversity. The process of mineralization performed by the soil microorganisms release fertilizer's nutrients into the environment, allowing arable crops to grow and develop properly. The ratio of the profit significantly depends on the kind of organic fertilizer. While fertilizers with low C/N ratio (slurries) release theirs nutrients rapidly and in a relatively huge proportion during the first year, fertilizers with high C/N ratio (farmyard manure) release theirs nutrients slowly but for a longer time period [33, 34]. The process of mineralization also depends on the climate conditions, decreasing significantly with the occurrence of dry periods. The main contribution of organic fertilizers, however, lies in the addition of organic matter to the soil [35], influencing soil chemical, physical, and biological properties.

Application of organic manure to the soil was a traditional way to maintain soil's fertility in the Czech Republic, especially to the 1990s, when animal husbandry and crop production was extensive. The agriculture sector was a priority for the communist leaders, and this era was so characterized with intensive application of mineral fertilizers (**Figure 1**) and organic manure (no statistical data are available up to 2007). After the Velvet Revolution in 1989, a significant decrease of mineral fertilizer consumption can be recorded, and this development continues in the case of P and K fertilizers. Nowadays, same doses of P and K fertilizers are applied on the arable land like in the 1960s. Situation is different in the case of mineral N, which is

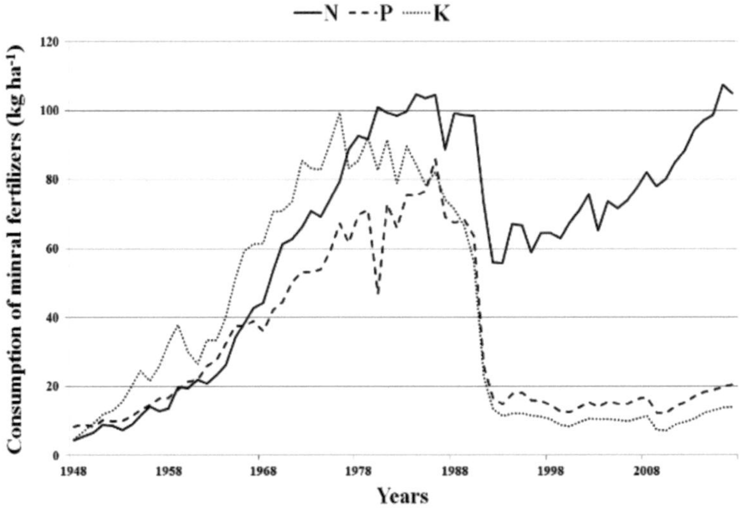

Figure 1.
Consumption of mineral N, P, and K (kg ha^{-1}) fertilizers in the Czech Republic from 1949 to 2017.

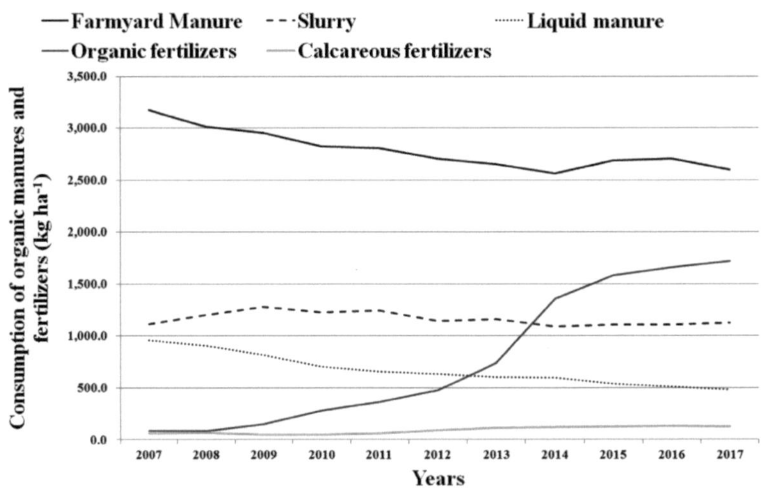

Figure 2.
Consumption of farmyard manure, slurry, liquid manure, organic fertilizers, and calcareous fertilizers in the Czech Republic from 2007 to 2017.

applied in approximately 100 kg ha^{-1}. Statistical data describing application of organic manure and fertilizers are available from 2007 (**Figure 2**). A long-term decrease trend can be observed in farmyard and liquid manure categories, which continues from the previous time era. A significant increase can be seen in the organic fertilizers category, which is connected with a huge boom of the biogas stations in the Czech Republic. However, digestates can serve as a source of nitrogen for the agriculture sector but cannot deal with the problem of organic matter in the soil.

The aforementioned information stand at the root of the problem of the present time when low doses of livestock manure (organic matter), together with reduced crop rotation, cause soil erosion and soil inability to support crops during extreme climate conditions, such as periods of droughts that will occur more frequently [36].

3. Materials and methods

3.1 Site description

In 1955, a series of three long-term crop rotation and fertilizer experiments was established on different soils (chernozems, cambisols) in the Czech Republic (**Figure 3**).

The most productive site is Ivanovice na Hané (ICRFE [37]); the intermediate is Čáslav (CCRFE [38]). The least productive is Lukavec (LCRFE [39]) crop rotation and fertilization experiment. Detailed descriptions of all experimental sites with climate description (precipitation and temperature, 2012–2015) are given in **Tables 1** and **2**.

Figure 3.
Ivanovice, Lukavec, and Čáslav experimental sites displayed on the map of the Czech Republic.

Parameter/locality	ICRFE	LCRFE	CCRFE
Altitude (m a.s.l.)	225	620	263
Soil type*	Chernozems leptic	Cambisols skeletic	Chernozems calcic (luvic)—degraded
Parent material	Loess, loess loam	Parabula metamorphosed	Loess, loess loam
Cropping area**	Sugar beet	Potato	Sugar beet
Thickness of the arable layer (cm)	40–45	25–30	30–35
Mean annual temperature (°C)***	9.1	7.4	8.3
Mean annual precipitation (mm)***	538.2	690.7	590.0

**WRB 2015.*
***According to the Czech national classification.*
****ICRFE weather station Ivanovice na Hané (1962–2012), LCRFE weather station Lukavec (1962–2012), CCRFE weather station Filipov (1982–2012).*

Table 1.
Basic description of Ivanovice (ICRFE), Čáslav (CCRFE), and Lukavec (LCRFE) crop rotation and fertilization experiments.

Locality	Year	Temperature (°C)		Precipitation (mm)	
		Mean annual	Mean annual in growing season	Annual	Annual in growing season
ICRFE*	2012	9.6	16.8	481.6	329.9
	2013	9.2	15.7	550.9	379.2
	2014	10.5	15.9	520.4	391.0
	2015	10.4	16.8	387.0	265.4
LCRFE**	2012	8.0	14.3	744.4	403.5
	2013	7.3	13.4	875.7	606.5
	2014	8.9	14.1	708.8	541.5
	2015	8.7	14.3	576.1	281.2
CCRFE***	2012	9.8	16.3	636.8	439.9
	2013	9.3	15.6	636.6	466.5
	2014	10.8	15.9	619.1	448.4
	2015	11.0	16.7	442.4	232.4

**ICRFE weather station Ivanovice na Hané.*
***LCRFE weather station Lukavec.*
****CCRFE weather station Filipov.*

Table 2.
Basic description of weather conditions (2012–2015) in Ivanovice (ICRFE), Čáslav (CCRFE), and Lukavec (LCRFE) crop rotation and fertilization experiments.

3.2 Experimental design

All experiments were established in the same standardized design in four field strips because four crops were in rotation. In each field strip, 12 fertilizer treatments were established in four replications arranged in completely randomized block design (12 × 4 = 48 experimental plots per field strip). The size of each experimental plot was 8 × 8 m, but only the central area 5 × 5 m was used for yield determination and soil sample collection. In each strip, two complete randomized blocks of all treatments were located (**Figure 4**) (a list of fertilizer treatments used in the experiment is given in **Tables 3** and **4**).

In this paper, only four of the most contrasting fertilizer treatments were analyzed: the control (Control), without any fertilizer input; the farmyard manure

a		b		c		d	
11	12	21	22	15	16	23	24
13	14	23	24	11	12	25	26
15	16	25	26	13	14	21	22
25	26	15	16	21	22	13	14
21	22	11	12	23	24	15	16
23	24	13	14	25	26	11	12

Figure 4.
Spatial arrangement of treatments in experimental strips; the same spatial arrangement was used in all four experimental strips. Letters (a–d) indicate complete randomized blocks, and Arabic numbers indicate individual treatments. Treatment numbers are given in ***Tables*** *3 and 4.*

Treatment	Nutrients applied by farmyard manure			Nutrients applied by mineral fertilizers			Total amount of applied nutrients			Distribution of N application
	N	P	K	N	P	K	N	P	K	
Control	0	0	0	0	0	0	0	0	0	
FYM + 0	8	4	12	0	0	0	8	4	12	
FYM + N_3PK	8	4	12	120	35	83	128	39	95	1, 2
FYM + N_2	8	4	12	80	0	0	88	4	12	1

Note: 1. Application of 80 kg N ha^{-1} in the spring before sowing; 2. application of 40 kg N ha^{-1} early in the spring (six leaves unfolded).

Table 3.
Nutrients applied directly to Zea mays *in kg ha^{-1} in all years of the experiment.*

Treatment	Nutrients applied by farmyard manure			Nutrients applied by mineral fertilizers			Total amount of applied nutrients		
	N	P	K	N	P	K	N	P	K
Control	0	0	0	0	0	0	0	0	0
FYM + 0	20	10	30	0	0	0	20	10	30
FYM + N_3PK	20	10	30	94	40	87	114	50	117
FYM + N_2	20	10	30	74	4	14	94	14	44

Table 4.
Mean annual application rates of nutrients (kg ha^{-1}) over all crops during the run of the experiment.

(FYM + 0) treatment; farmyard manure together with mineral N, P, and K, fertilizers (FYM + N_3PK); and farmyard manure together with mineral N (FYM + N_2).

Except for the Control, the FYM application was performed in the autumn before the preceding root crop planting. Mineral P and K fertilizers and FYM were plowed down immediately after application. The distribution of mineral N application within the year was the following: (1) application of 80 kg N ha^{-1} in the spring before sowing and (2) application of 40 kg N ha^{-1} early in the spring (six leaves unfolded) (**Table 3**). A 4-year crop rotation system was used during the run of the experiment: maize, spring barley, oilseed rape, and winter wheat. Straw of cereals and residues of other crops were removed from the experimental plots after the harvest of the main product. Pesticides have been applied if necessary, and growth regulators have never been used. Average yield (dry mass of silage maize) values of main products during the 2012–2015 are given in **Figure 5**.

3.3 Sampling and soil analysis

Sampling from the upper Ap horizon was done in the 2012–2015 period. Each treatment (in selected year) was sampled four times (n = 4), altogether 64 samples for the studied period. Soil reaction was determined by potentiometric method in 50 ml of 0.2 mol KCl (inoLab pH 730, WTW, Germany). The SOC (Corg) content was analyzed colorimetrically according to Sims and Haby [40] and also by oxidimetric titration, according to Nelson and Sommers [41]. Total nitrogen content was determined with concentrated sulfuric acid in a heating block (Tecator, Sweden), followed by the Kjeldahl method [42, 43]. Up to discovering the Mehlich III method [44], the concentrations of P, K, and Mg were analyzed by the Mehlich II and Mehlich I methods. Concentrations of P, K, and Mg were then analyzed by ICP-OES (Thermo Scientific iCAP 7400 Duo, Thermo Fisher Scientific, Cambridge, UK).

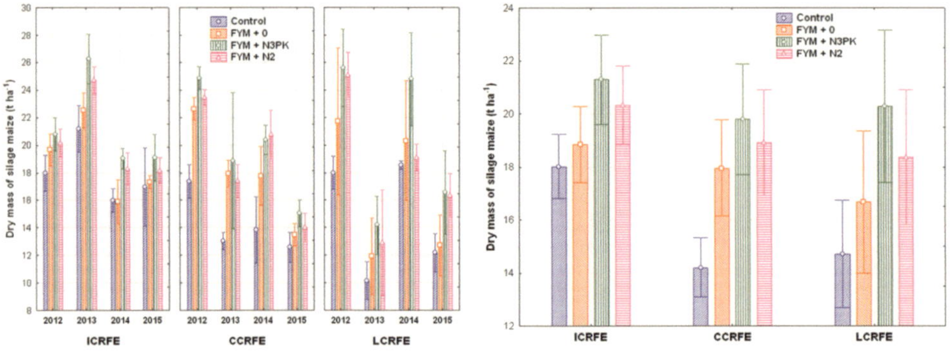

Figure 5.
The yield dry mass of silage maize and the average yield in period 2012–2015.

3.4 Data analysis

Statistical analysis, including graphical outputs, was carried out using Statistica 13 (TIBCO Software Inc., Palo Alto, USA, 2018). For the statistical data processing and evaluation we applied exploratory data analysis (EDA), analysis of variance (ANOVA), Tukey test (HSD test), Fisher's LSD test (LSD test), linear regression (LR), principal component analysis (PCA), factor analysis (FA), and cluster analysis (CLU). LR was calculated by the QC.Expert 3.3^{Pro} statistical program (TriloByte Statistical Software, Ltd., Pardubice, CZ, 2018). The linear regression diagnostics was solved with the aid of a technique called regression triplet [45]. PCA was used for interpreting the parameters of soil organic matter (C_{org}, C/N ratio, etc.) and physicochemical properties of soil (pH, content of nitrogen, phosphorus, calcium, potassium, magnesium, etc.). Selected measured characteristics were used as predictors (factors); they were chosen on the basis of an eigenvalue graph. Variables with the impaired assumption of normality were converted using logarithmic transformation. As a part of step 1, PCA was carried out with all the variables to calculate the most important variables. Step 2 involved selecting active and supplementary variables for better interpretation. In the case of a lower number of samples, this stepwise analysis significantly improves the outcome of the PCA. The PCA was used for calculating a component weight for the investigated variables. Based on correlations and contributions in convincing factors, each of the characteristics was subsequently assessed for relevance explaining the multidimensional dependencies (correlations) in the factorial plane. The factor analysis (FA) analyzed the internal contexts and relationships (correlations) and revealed the basic structure of the source data matrix. The FA also identified factors and then assigned to each factor a content meaning (physical or chemical) [45].

The CLU was used for classification of objects to the clusters. The CLU does not differentiate significant and insignificant markers but differentiate the significant clusters [45]. The CLU was performed by a complete linkage method. The statistical significance was assessed at a significance level of $p = 0.05$.

4. Results

4.1 Carbon, nitrogen, C/N ratio (SOM)

The SOC content (C_{org}) ranged in individual years (2012–2015) at all three sites from 1.05 to 2.38% (**Figure 6**). Higher SOC contents were recorded at ICRFE and

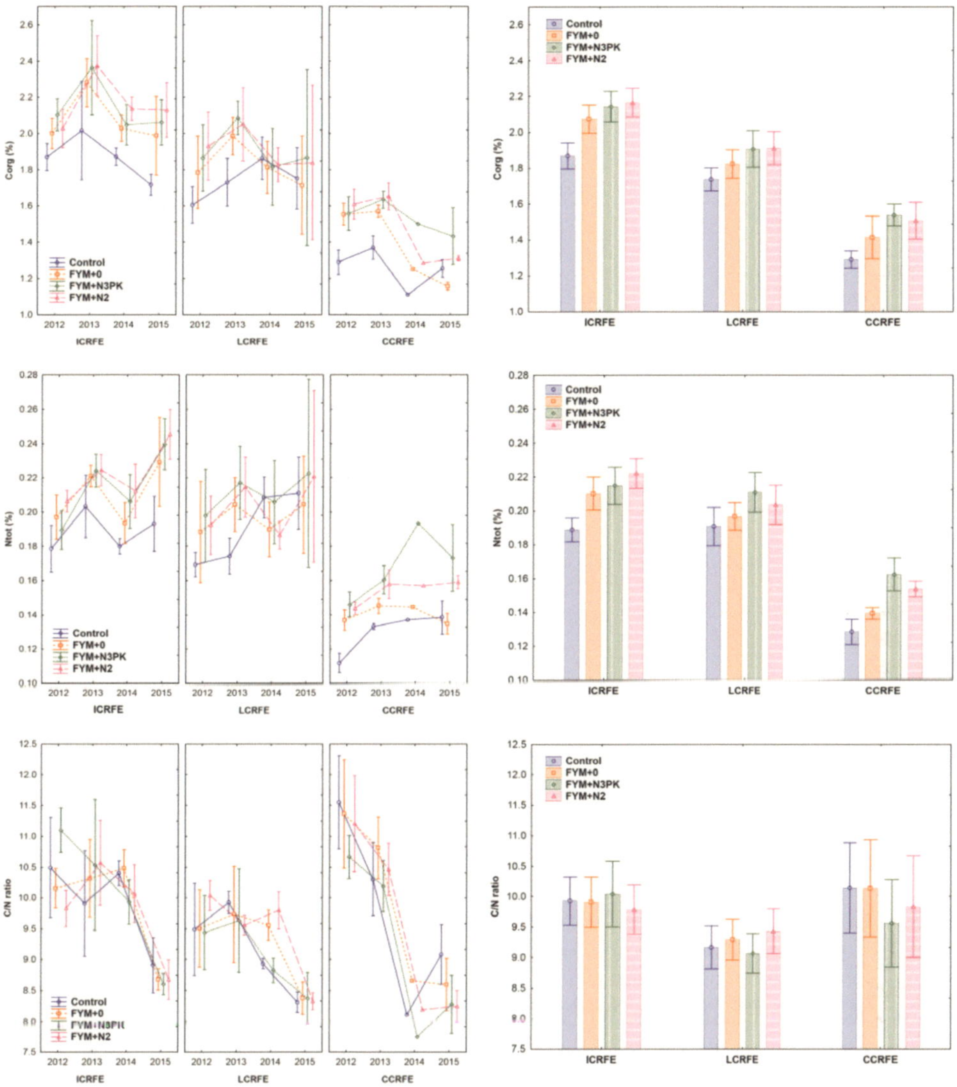

Figure 6.
Soil organic carbon (C_{org}) content, total nitrogen (N_{tot}), and C/N ratio as affected by fertilizer treatment and locality during studied period (2012–2015).

LCRFE, compared to CCRFE (degraded chernozem). Statistically higher SOC content (HSD test) in individual years and for the 2012–2015 period was recorded in FYM + N_3PK and FYM + N_2 treatments, compared to the Control treatment at all three sites and FYM + N_2 treatment (LCRFE and CCRFE). Similar results, including statistically significant differences (HSD test), were also recorded for the nitrogen content (N_{tot}) in individual years and in 2015–2015 period. The concentration of N_{tot} ranged from 0.11 to 0.25% (**Figure 6**). The average C/N ratio ranged from 9.5 to 10.1 at ICRFE (2012–2015, without significant differences between the treatments) and CCRFE (significantly lower C/N ratio in the FYM + N_3PK treatment). Lower C/N ratio was observed at LCRFE (without statistically significant difference between the treatments), where the ratio ranged from 9.0 to 9.5 (**Figure 6**). Statistically significant linear regression (data from all three sites) between the SOC and N_{tot} was recorded (equation parameters for N_{tot} (%): y = 0.0293 + 0.0886 * C_{org}; R = 0.8139; p = 0.0000; R^2 = 0.6624; mean quadratic error of prediction (MEP) = 0.0003; Akaike information criteria (AIC) = −4163.4912 (**Figure 7**)).

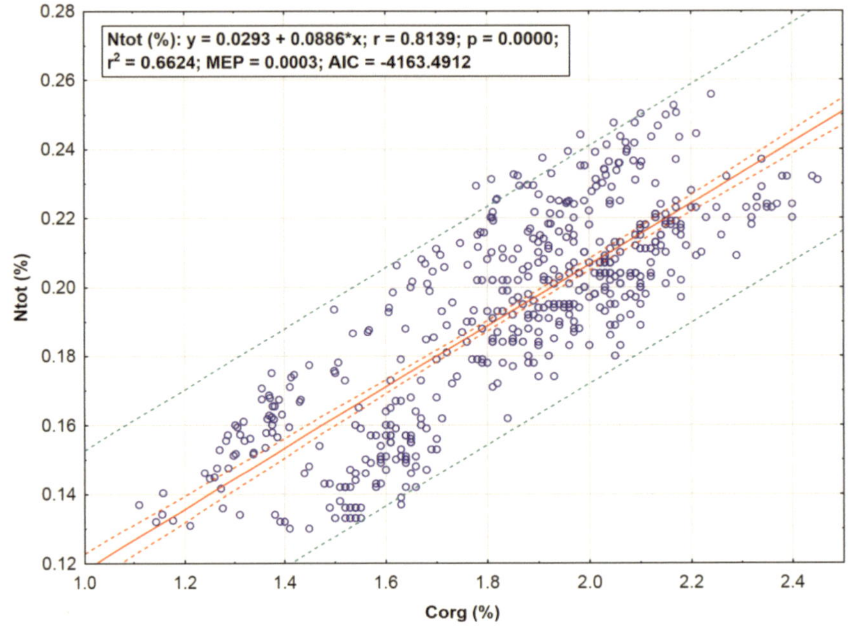

Figure 7.
The linear regression dependence of the N_{tot} on C_{org} during the studied period (2012–2015).

4.2 Soil reaction and nutrients

The value of pH at ICRFE and CCRFE (chernozems and degraded chernozems) ranged from 6.5 to 7.4 (**Figure 8**). The decrease of the pH was recorded at both localities and both treatments (FYM + N_3PK and FYM + N_2), without any statistically significant difference in particular years and during the whole 2012–2015 time period (**Figure 8**). The lower pH value was recorded at LCRFE (cambisol). At this site (LCRFE), a significantly lower pH value was recorded in the FYM + N_3PK treatment (2012–2015), compared to other fertilizer treatments.

The contents of plant available P and K at all three experimental sites and in all fertilizer treatments in individual years showed a similar trend (except of the FYM + N_3PK treatment at LCRFE, where a gradual decrease of P was recorded). The average P content ranged from 45 to 195 mg kg^{-1} (2012–2015). The highest content, at all sites, was recorded in the FYM + N_3PK treatment (significantly higher than in the Control and FYM + 0 treatments, **Figure 9**). Statistically

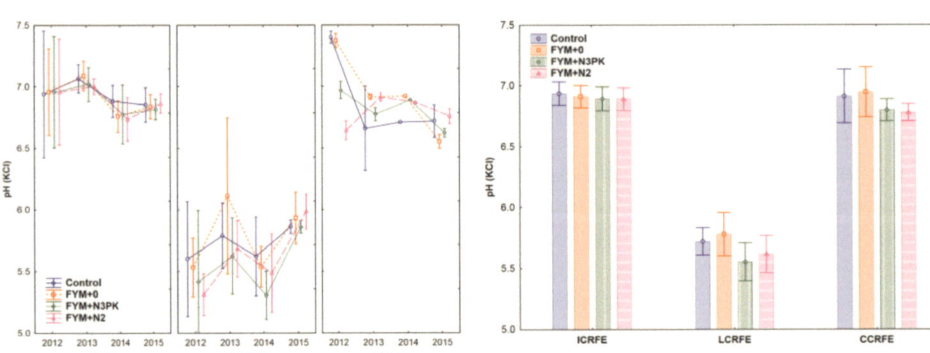

Figure 8.
The pH value (KCl) as affected by fertilizer treatment and locality during the studied period (2012–2015).

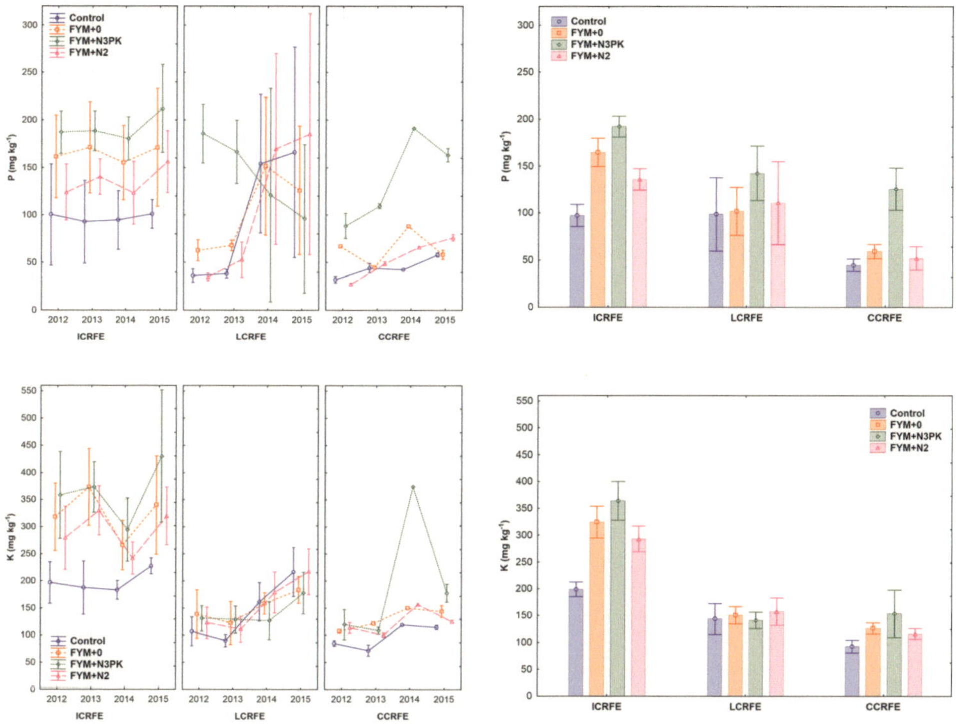

Figure 9.
Soil available macronutrients (P, K/mg kg^{-1}/) as affected by fertilizer treatment and locality during the studied period (2012–2015).

significant linear regression dependency (data from all three sites) between the K and P content was recorded (parameters of K equation: (mg kg^{-1}): y = 69.8464 + 1.0883 * P (mg kg^{-1}); R = 0.7331; p = 0.0000; R^2 = 0.5375; MEP = 3154.2151; AIC = 3674.2704 (**Figure 10**)).

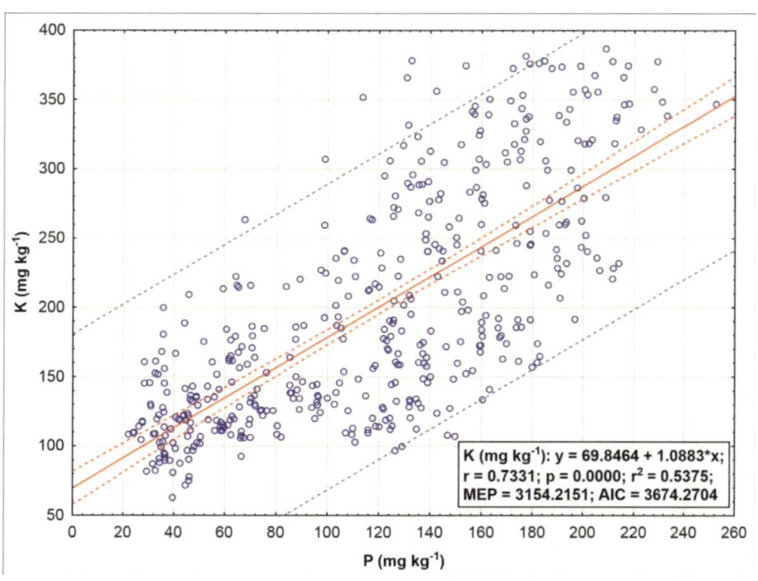

Figure 10.
The relationship between K (mg kg^{-1}) and P (mg kg^{-1}) contents during the studied period (2012–2015).

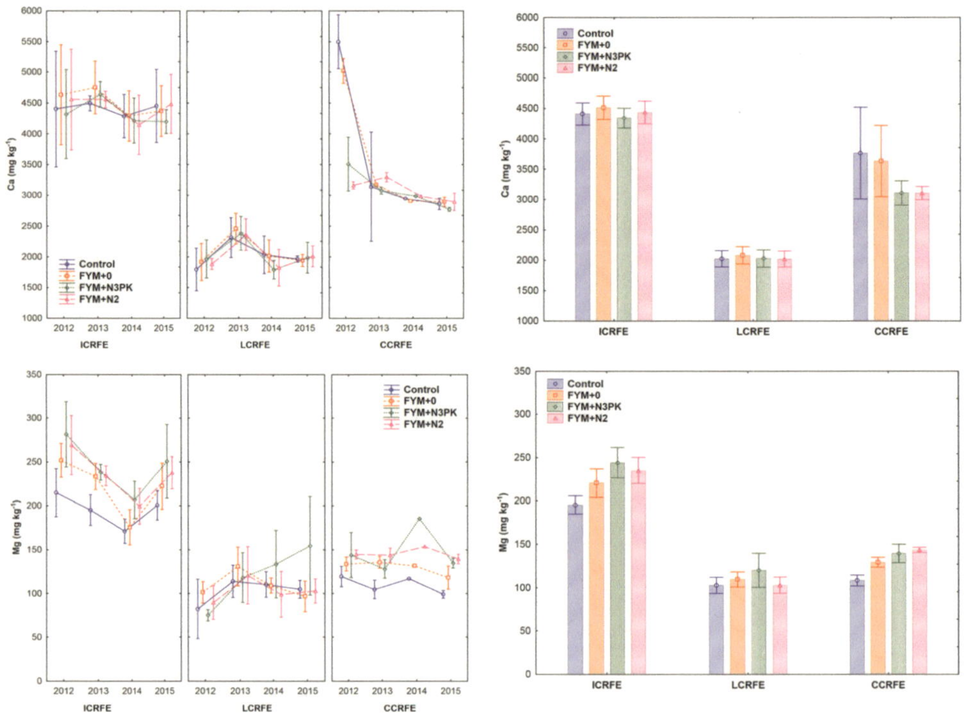

Figure 11.
Soil available macronutrients (Ca, Mg (mg kg^{-1})) as affected by fertilizer treatment and locality during the studied period (2012–2015).

The contents of available Ca and K at all three sites and in all fertilizer treatments in individual years showed a similar trend (**Figure 11**). The Ca content at ICRFE and LCRFE sites is balanced and without significant differences between the treatments. At all three sites, higher levels were recorded in the Control and FYM + 0 treatments (significantly higher at CCRFE). The average Mg content for the 2012–2015 period ranged from 100 to 250 mg kg^{-1} (higher content at ICRFE, lower contents at LCRFE and CCRFE). The highest contents were recorded in the FYM + N_3PK treatment at all experimental sites (significantly higher compared to the Control and FYM + 0 treatments).

4.3 Multi-criteria evaluation of measured soil attributes and parameters

According to the eigenvalue results, the first two axes are significant on the PC1 and PC2 component figure (PCA), which together represent about 95% of the variability (2012–2015, **Figure 12**).

The PC1 axis in **Figure 12** (PC1 × PC2) represents the content of available nutrients (K, Mg, P) and the SOM (C_{org}). The K content is strongly and negatively correlated with this axis ($r = -0.98$). Similar correlations are in the case of Mg content ($r = -0.91$), C_{org} ($r = -0.88$), P ($r = -0.88$), P ($r = -0.86$), and N_{tot} ($r = -0.77$). The PC2 axis represents correlation with the pH value ($r = 0.95$) and Ca content ($r = 0.78$). According to the projection of the cases (**Figure 12**), the fertilizer treatments and localities are separated (clusters close together that behave similarly are correlated). According to the analysis, the Control treatment at ICRFE is significantly separated from other treatments. At LCRFE and CCRFE sites, two clusters are separated: (1) Control and FYM + 0 and (2) FYM + N_3PK with

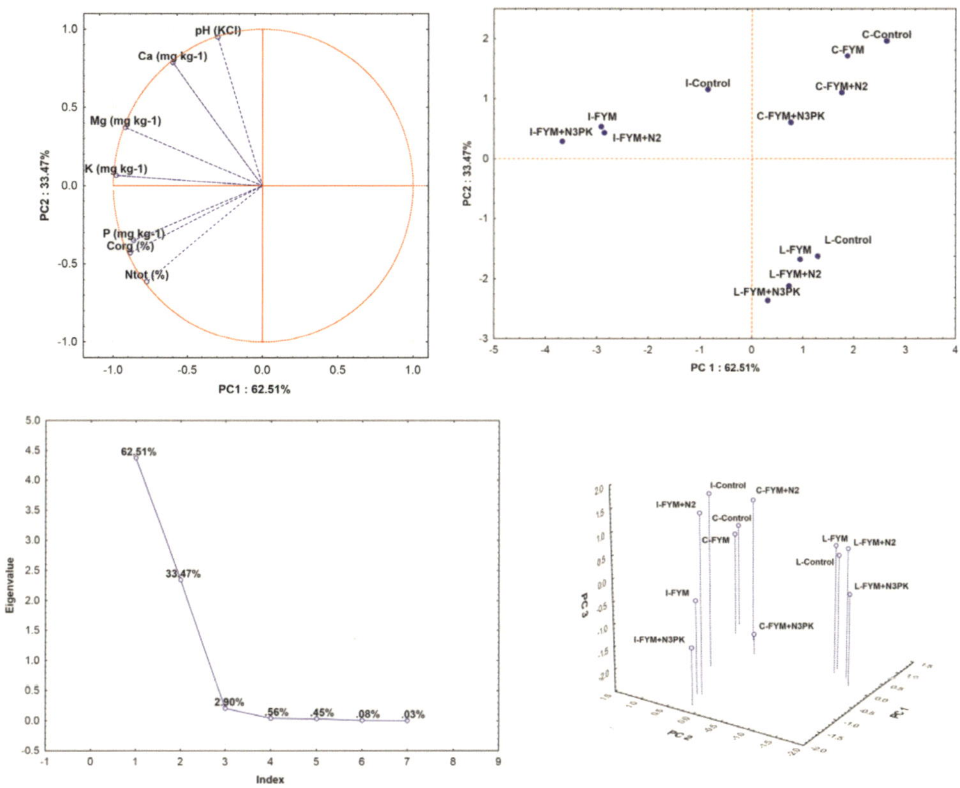

Figure 12.
The PCA of studied parameters of soil organic matter (C_{org}, N_{tot}), soil available macronutrients (P, K, Ca, Mg), and soil reaction (pH) during the studied period (2012–2015). Note: I, ICRFE; L, LCRFE; C, CCRFE.

FYM + N_2. Factor 1 in the FA (**Figure 13**) describes the properties from the point of view of the SOC (decomposition processes) and nutrient content. Factor 2 describes the soil from the point of view of pH value.

The communality represents the proportion of variability of attributes expressed by the factors involved. It is similar to R^2 value we get when explaining the original characters by regression of selected factors [45]. From the contribution of factors 1 and 2 to the communality, it's clear how communality acquires high values (more than 0.9), and thus, the values of attributes are precisely considered by the proposed factor model (**Table 5**).

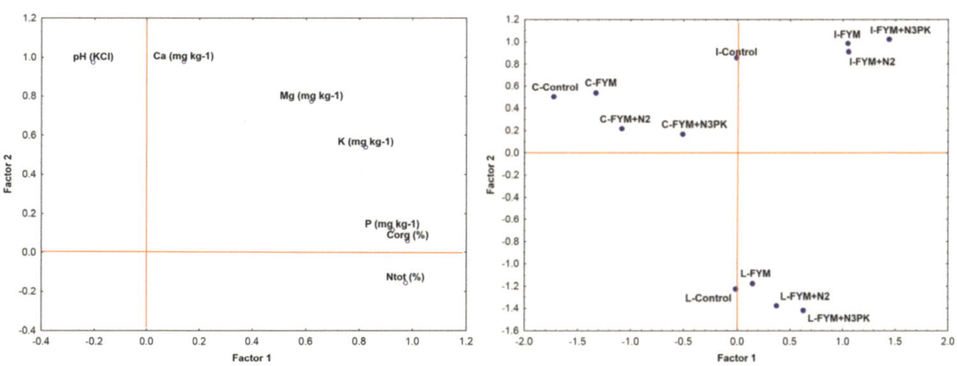

Figure 13.
The FA of studied parameters of soil organic matter (C_{org}, N_{tot}), soil available macronutrients (P, K, Ca, Mg), and soil reaction (pH) during the studied period (2012–2015). Note: I, ICRFE; L, LCRFE; C, CCRFE.

Parameter	Factor weights		Factors contribution		
	Factor 1	Factor 2	Factor 1	Factor 2	Communality
pH (KCl)	−0.2063	0.9737	0.0426	0.9906	0.9876
C_{org} (%)	0.9818	0.0578	0.9638	0.9672	0.9966
N_{tot} (%)	0.9736	−0.1564	0.9479	0.9724	0.9962
P (mg kg^{-1})	0.9241	0.1151	0.8540	0.8673	0.9126
K (mg kg^{-1})	0.8232	0.5375	0.6777	0.9665	0.9718
Ca (mg kg^{-1})	0.1389	0.9770	0.0193	0.9739	0.9630
Mg (mg kg^{-1})	0.6200	0.7721	0.3844	0.9806	0.9865

Table 5.
The factor weights and contributions of selected factors to the communality for each parameters in factor analysis (FA).

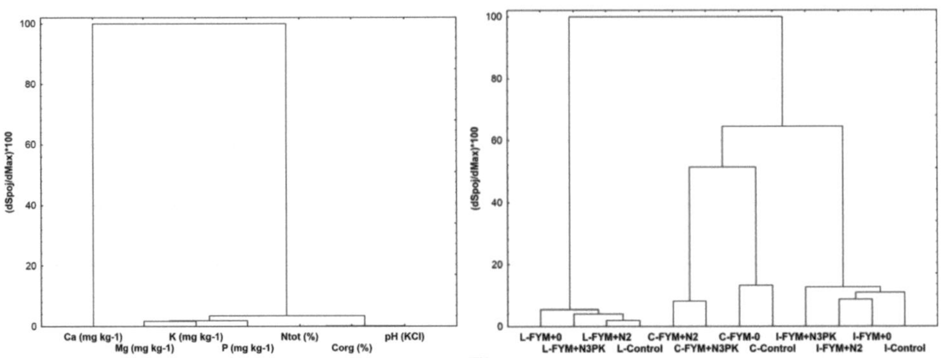

Figure 14.
CLU of studied parameters of soil organic matter (C_{ox}, N_{tot}), soil available macronutrients (P, K, Ca, Mg), and soil reaction (pH) during the studied period (2012–2015). Note: I, ICRFE; L, LCRFE; C, CCRFE.

The dendrograms (**Figure 14**) accomplished by complete linkage method prove separated clusters of the Control treatment and inosculated clusters of FYM + 0, FYM + N_3PK, and FYM + N_2 fertilizer treatments (ICRFE and LCRFE). Two clusters were recorded at CCRFE site—Control with FYM + 0 and FYM + N_3PK with FYM + N_2. Similarly, the CLU analysis also divided the soil properties—a cluster of nutrient content, SOM, and the pH value. The results of the CLU are consistent with the results of FA and PCA.

5. Discussion

The results (2012–2015) from the long-term fertilizer experiments, established in 1955 in different soil-climate conditions (Ivanovice na Hané, Lukavec, and Čáslav) in the Czech Republic (representation of soil types in the Czech Republic: chernoszems 11%, luvisols 12%, cambisols 46% [46]), proved that the application of the farmyard manure (FYM) increases the SOM (C_{org}) content in the soil (**Figure 1**). If we apply the FYM together with mineral fertilizers (NPK), the conditions of the soil and its quality are maintained at optimum quality (still increasing the C_{org} content in the soil (**Figure 6**)). These results are confirmed by the study of Zhao et al. [47], performed under similar conditions (the average annual air temperature and precipitation are in the range of 8–15°C and 500–950 mm) in the North China Plain. Benefits of the long-term application of organic manure and mineral fertilizers on SOC pools and sequestration also confirmed studies from Liu et al. [48] and Menšík et al. [14], who analyzed the effect of the long-term application of organic manures and NPK on soil quality parameters (brown soils) in the Czech Republic. The application of organic manures and slurries to the soil increases soil fertility and ensures stable production and food security for future generations, even under changing environmental conditions. On the other hand, application of mineral fertilizers without any inputs of organic matter leads to destabilization of soil environment and is connected with a significant decrease of soil fertility. This is confirmed by studies of Yang et al. [49] and Chen et al. [50], who published that application of mineral fertilizers is insufficient for maintaining the SOC under the conditions of conventional agriculture, where no plant residues or manures are returned to the soil.

Increasing C_{org} content is closely connected with increasing N_{tot} (see linear relationship between C_{org} and N_{tot} (**Figures 6** and 7)). The same results were published by Maltas et al. [51] from Switzerland, who analyzed the effect of the cattle manure and NPK in the long-term experiment, established in 1976, on calcaric cambisols.

Nitrogen fertilization has been reported to increase C sequestration [52, 53], but the effect differs between the studies [51, 54, 55]. The application of fertilizers to nutrient-deficient soils generally increases the SOC content because fertilizers increase the crop production and, thereby, the amount of plant residues released to the soil [56]. Thus, the strategy in soil nitrogen management is continuous incorporation of organic manure, to prevent acidification and maintain high soil productivity [57–59].

According to our results, application of mineral fertilizers and organic manures did not significantly influence phosphorus, potassium, and magnesium content during the studied period (2012–2015). Results from the Maltas et al. [51] also show that N fertilization significantly reduces the available Mg content in the soil, where nitrogen fertilization likely increased Mg extraction from the soil, due to higher crop yields [60]. This has not been proved in our study. On the contrary, the Mg content in the soil increased (**Figure 11**). The highest N, P, and K content in top soil was recorded in FYM + N_3PK treatments, while the lowest content was in the Control treatment. Our results are consistent with the study of Yang et al. [31]. He showed that application of organic manure in cultivated farmland significantly increased nitrogen accumulation rate, and this influence was greater in comparison with NPK application.

The multivariate exploratory techniques help us to determine the structure and interrelationships between objects and attributes by the technique of the reduction of the attributes on the latent variable [45]. The aim of the PCA is to simplify the description of a group of mutually dependent or correlated attributes [61], while the FA serves to examine relationships and correlations between a large number of original attributes by using a set of less latent variables, called factors [62]. The multivariate exploratory techniques (PCA, FA, and CLU) divided different techniques of ecosystems into two categories (**Figures 12–14**): (1) Control and FYM + 0: higher C_{org}, N_{tot}, and pH value. The contents of P, K, Ca, and Mg were comparable to the Control treatment (except of ICRFE, where chernozems are occurring). (2) FYM + N_3PK; FYM + N_2 treatments: higher C_{org} and N_{tot} content, higher content of plant available P, K, and Mg, especially in FYM + N_3PK treatment, and higher pH value (a decrease compared to the Control and FYM + 0 treatments —acidification due to mineral nitrogen application on soils with worse properties in LCRFE and CCRFE sites (**Figures 12** and **13**)).

6. Conclusion

More than 60 years of continuous fertilization with organic manures and mineral NPK fertilizers on different soil types (chernozems, degraded chernozems, and cambisols) and in different climatic conditions in the Czech Republic (the Central Europe) leads to a significant differentiation of soil properties in terms of soil quality. According to the multi-criterial evaluation (PCA, FA, CLU), we separated three different soil-climate localities (SOC and N_{tot} content, the C/N ratio, soil acidity and content of available nutrients) and different fertilizer treatments (Control, FYM + 0, FYM + N_3PK, and FYM + N_2) in each locality.

The Control and FYM + 0 treatments (the basic soil conditions) are characterized by a higher content of C_{org}, N_{tot}, and pH value. The P, K, Ca, and Mg content in the FYM + 0 treatments was comparable to the Control (except of chernozem soil type—ICRFE). The FYM + N_3PK and FYM + N treatments are characterized by a high content of C_{org} and N_{tot}; higher content of available nutrients (P, K, and Mg), especially in FYM + N_3PK treatment; and slight decrease of the pH value (compared to the Control and FYM + 0 treatments—acidification of the soil due to the application of N in mineral form, especially in wore conditions of LCRFE and CCRFE).

The long-term application of organic manures, and organic manures with mineral NPK (or N), maintains the soil in optimal quality (soil fertility), stabilizes the production in terms of quantity and quality of food and feedstuff, and increases the adaptive potential of current land to the changing environmental conditions.

The multivariate exploratory techniques, such as PCA, FA, and CLU, are very suitable methods for displaying, evaluating, and interpreting the data and results about the physicochemical soil properties.

Acknowledgements

This paper is supported by the Ministry of Agriculture of the Czech Republic project ČR-RO0418 and by the Czech National Agency for Agricultural Research (NAZV)—project QK1810010 and QK1910334.

Author details

Ladislav Menšík*, Lukáš Hlisnikovský and Eva Kunzová
Division of Crop Management Systems, Crop Research Institute, Praha, Czech Republic

*Address all correspondence to: ladislav.mensik@vurv.cz

References

[1] IPCC. IPCC special report on renewable energy sources and climate change mitigation. Prepared by Working Group III of the Intergovernmental Panel on Climate Change. Cambridge, United Kingdom and New York, NY, USA: Cambridge University Press; 2011

[2] IPCC. Climate change 2014: Mitigation of climate change. Contribution of Working Group III to the Fifth Assessment Report of the Intergovernmental Panel on Climate Change. Cambridge, United Kingdom and New York, NY, USA: Cambridge University Press; 2014

[3] Tilman D, Cassman KG, Matson PA, Naylor R, Polasky S. Agricultural sustainability and intensive production practices. Nature. 2002;**418**:671-677

[4] Godfray HCJ, Beddington JR, Crute IR, et al. Food security: The challenge of feeding 9 billion people. Science. 2010; **327**:812-818

[5] Urruty N, Tailliez-lefebvre D, Huyghe C, Tailliez-lefebvre D. Stability, robustness, vulnerability and resilience of agricultural systems. A review. Agronomy for Sustainable Development. 2016;**36**:15

[6] Lobell DB, Schlenker W, Costa-Roberts J. Climate trends and global crop production since 1980. Science. 2011;**333**:616-620

[7] Jarvis A, Lau C, Cook S, et al. An integrated adaptation and mitigation framework for developing agricultural research: Synergies and trade-offs. Experimental Agriculture. 2011;**47**: 185-203

[8] Knox J, Hess T, Daccache A, Wheeler T. Climate change impacts on crop productivity in Africa and South Asia. Environmental Research Letters. 2012;**7**: 041001

[9] Rosenzweig C, Elliott J, Deryng D, et al. Assessing agricultural risks of climate change in the 21st century in a global gridded crop model intercomparison. Proceedings of the National Academy of Sciences. 2014;**111**: 3268-3273

[10] Steenwerth KL, Reynolds MP, Sandoval Solis S, et al. Climate-smart agriculture global research agenda: Scientific basis for action [electronic resource]. Agriculture and Food Security. 2014;**3**:56

[11] Lone B, Qayoom S, Singh P, et al. Climate change and its impact on crop productivity. British Journal of Applied Science and Technology. 2017;**21**:1-15

[12] Blanco M, Ramos F, Van Doorslaer B, et al. Climate change impacts on EU agriculture: A regionalized perspective taking into account market-driven adjustments. Agricultural Systems. 2017;**156**:52-66

[13] Frank S, Witzke HP, Zimmermann A, Havlík P, Ciaian P. Climate change impacts on european agriculture: A multi model perspective. In: EAAE 2014 Congress Agri-Food and Rural Innovations for Healthier Societes; 26–29 August 2014; Ljublana, Slovenia. 2014

[14] Menšík L, Hlisnikovský L, Pospíšilová L, Kunzová E. The effect of application of organic manures and mineral fertilizers on the state of soil organic matter and nutrients in the long-term field experiment. Journal of Soils and Sediments. 2018;**18**:2813-2822

[15] Banwart S, Black H, Cai Z, et al. The global challenge for soil carbon. In: Banwart S, Noellemeyer E, Milne E,

editors. Soil Carbon: Science, Management and Policy for Multiple Benefits. Oxfordshire: CAB International; 2015. pp. 1-9

[16] Stevenson FJ. Humus Chemistry: Genesis, Composition, Reactions. New York: John Wiley and Sons; 1994

[17] Piccolo A. Humus and soil conservation. In: Piccolo A, editor. Humic Substances in Terrestrial Ecosystems. Amsterdam: Elsevier Science B.V.; 1996. pp. 225-264

[18] Meng Q, Sun Y, Zhao J, et al. Distribution of carbon and nitrogen in water-stable aggregates and soil stability under long-term manure application in solonetzic soils of the Songnen plain, Northeast China. Journal of Soils and Sediments. 2014;**14**:1041-1049

[19] Pospíšilová L, Hábová M, Vlček V, Jandák J, Menšík L, Barančíková G. Carbon dynamic after conversion of permanent grassland into arable soil. Agriculturae Conspectus Scientificus. 2018;**83**:25-30

[20] Smith J, Smith P, Wattenbach M, et al. Projected changes in the organic carbon stocks of cropland mineral soils of European Russia and the Ukraine, 1990–2070. Global Change Biology. 2007;**13**:342-356

[21] Crowther T, Todd-Brown K, Rowe C, et al. Quantifying global soil C losses in response to warming. Nature. 2016; **104**:104-108

[22] Lal R. Soil carbon stocks under present and future climate with specific reference to European ecoregions. Nutrient Cycling in Agroecosystems. 2008;**81**:113-127

[23] Qin S, Hu C, He X, Dong W, Cui J, Wang Y. Soil organic carbon, nutrients and relevant enzyme activities in particle-size fractions under conservational versus traditional agricultural management. Applied Soil Ecology. 2010;**45**:152-159

[24] King JA, Bradley RI, Harrison R. Current trends of soil organic carbon in English arable soils. Soil Use and Management. 2005;**21**:189-195

[25] Lal R. Enhancing crop yields in the developing countries through restoration of the soil organic carbon pool in agricultural lands. Land Degradation and Development. 2006;**17**: 197-209

[26] Plaza-Bonilla D, Arrúe JL, Cantero-Martínez C, Fanlo R, Iglesias A, Álvaro-Fuentes J. Carbon management in dryland agricultural systems. A review. Agronomy for Sustainable Development. 2015;**35**:1319-1334

[27] Kubát J, Lipavský J. Evaluation of organic matter content in arable soils in the Czech Republic. In: Behl RK, Merbach W, Meliczek H, Kaetsch C, editors. Crop Science and Land Use for Food and Bioenergy. Jodhpur, India: Agrobios (International); 2010. pp. 245-251

[28] CZSO. Statistical Yearbook of the Czech Republic—2016. Praha: Czech Statistical Office; 2017

[29] Zhang W, Xu M, Wang X, et al. Effects of organic amendments on soil carbon sequestration in paddy fields of subtropical China. Journal of Soils and Sediments. 2012;**12**:457-470

[30] Ren T, Wang J, Chen Q, Zhang F, Lu S. The effects of manure and nitrogen fertilizer applications on soil organic carbon and nitrogen in a high-input cropping system. PLoS One. 2014; **9**(5):e97732

[31] Yang R, Su YZ, Wang T, Yang Q. Effect of chemical and organic fertilization on soil carbon and nitrogen accumulation in a newly cultivated

farmland. Journal of Integrative Agriculture. 2016;**15**:658-666

[32] Verma G, Sharma RP, Sharma SP, Subehia SK, Shambhavi S. Changes in soil fertility status of maize-wheat system due to long-term use of chemical fertilizers and amendments in an alfisol. Plant, Soil and Environment. 2012;**58**: 529-533

[33] Bhogal A, Williams JR, Nicholson FA, Chadwick DR, Chambers KH, Chambers BJ. Mineralization of organic nitrogen from farm manure applications. Soil Use and Management. 2016;**32**:32-43

[34] Ling LL, Tian LS. Nitrogen mineralization from animal manures and its relation to organic N fractions. Journal of Integrative Agriculture. 2014; **13**:2040-2048

[35] Liang Q, Chen H, Gong Y, Yang H, Fan M, Kuzyakov Y. Effects of 15 years of manure and mineral fertilizers on enzyme activities in particle-size fractions in a North China plain soil. European Journal of Soil Biology. 2014; **60**:112-119

[36] Olesen JE, Trnka M, Kersebaum KC, et al. Impacts and adaptation of European crop production systems to climate change. European Journal of Agronomy. 2011;**34**:96-112

[37] Kunzová E, Hejcman M. Yield development of winter wheat over 50 years of FYM, N, P and K fertilizer application on black earth soil in the Czech Republic. Field Crops Research. 2009;**111**:226-234

[38] Kunzová E, Hejcman M. Yield development of winter wheat over 50 years of nitrogen, phosphorus and potassium application on greyic Phaeozem in the Czech Republic. European Journal of Agronomy. 2010; **33**:166-174

[39] Hejcman M, Kunzová E. Sustainability of winter wheat production on sandy-loamy Cambisol in the Czech Republic: Results from a long-term fertilizer and crop rotation experiment. Field Crops Research. 2010; **115**:191-199

[40] Sims JR, Haby VA. Simplified colorimetric determination of soil organic matter. Soil Science. 1971;**112**: 137-141

[41] Nelson DW, Sommers LE. Total carbon, organic carbon, and organic matter. In: Sparks DL et al., editors. Methods of Soil Analysis. Vol. Part 3. 1996. pp. 961-1010

[42] Kjeldahl JGCT. Neue methode zur estimmung des stickstoffs in organischen korpern. Fresenius' Journal of Analytical Chemistry. 1883;**22**: 366-382. Available from: http://dx.doi.org/10.1007/BF01338151

[43] Kirk PL. Kjeldahl method for total nitrogen. Analytical Chemistry. 1950;**22**: 354-358

[44] Mehlich A. Mehlich 3 soil test extractant: A modification of Mehlich 2 extractant. Communications in Soil Science and Plant Analysis. 1984;**15**: 1409-1416

[45] Meloun M, Militký J. Statistical Data Analysis: A Practical Guide with 1250 Exercises and Answer Key on CD. New Delhi: Woodhead Publishing India; 2011

[46] Němeček J, Mühlhanselová M, Macků J, Vavříček D, Novák P. Taxonomický klasifikační systém půd. Praha: ČZU v Praze; 2011

[47] Zhao X, Hu K, Stahr K. Simulation of SOC content and storage under different irrigation, fertilization and tillage conditions using EPIC model in the North China Plain. Soil and Tillage Research. 2013;**130**:128-135

[48] Liu E, Yan C, Mei X, Zhang Y, Fan T. Long-term effect of manure and fertilizer on soil organic carbon pools in dryland farming in Northwest China. PLoS One. 2013;**8**(2):e56536

[49] Yang X, Zhang X, Fang H, Zhu P, Ren J, Wang L. Long-term effects of fertilization on soil organic carbon changes in continuous corn of northeast china: RothC model simulations. Environmental Management. 2003;**32**: 459-465

[50] Chen Y, Zhang X, He H, et al. Carbon and nitrogen pools in different aggregates of a Chinese Mollisol as influenced by long-term fertilization. Journal of Soils and Sediments. 2010;**10**: 1018-1026

[51] Maltas A, Kebli H, Oberholzer HR, Weisskopf P, Sinaj S. The effects of organic and mineral fertilizers on carbon sequestration, soil properties, and crop yields from a long-term field experiment under a Swiss conventional farming system. Land Degradation and Development. 2018;**29**:926-938

[52] Lemke RL, VandenBygaart AJ, Campbell CA, Lafond GP, Grant B. Crop residue removal and fertilizer N: Effects on soil organic carbon in a long-term crop rotation experiment on a Udic Boroll. Agriculture, Ecosystems and Environment. 2010;**135**:42-51

[53] Liebig MA, Varvel GE, Doran JW, Wienhold BJ. Crop sequence and nitrogen fertilization effects on soil properties in the Western Corn Belt. Soil Science Society of America Journal. 2014;**66**:596-601

[54] Blanchet G, Gavazov K, Bragazza L, Sinaj S. Responses of soil properties and crop yields to different inorganic and organic amendments in a Swiss conventional farming system. Agriculture, Ecosystems and Environment. 2016;**230**:116-126

[55] Khan SA, Mulvaney RL, Ellsworth TR, Boast CW. The myth of nitrogen fertilization for soil carbon sequestration. Journal of Environment Quality. 2007;**36**:1821-1832

[56] Edmeades DC. The long-term effects of manures and fertilisers on soil productivity and quality: A review. Nutrient Cycling in Agroecosystems. 2003;**66**:165-180

[57] Cai Z, Wang B, Xu M, Zhang H, Zhang L, Gao S. Nitrification and acidification from urea application in red soil (Ferralic Cambisol) after different long-term fertilization treatments. Journal of Soils and Sediments. 2014;**14**:1526-1536

[58] Cai Z, Wang B, Xu M, et al. Intensified soil acidification from chemical N fertilization and prevention by manure in an 18-year field experiment in the red soil of southern China. Journal of Soils and Sediments. 2015;**15**:260-270

[59] Yong SR, Dong LZ, Li Y, et al. Mechanisms for increasing soil resistance to acidification by long-term manure application. Soil and Tillage Research. 2019;**185**:77-84

[60] Sinaj S, Richner W, Flisch R, Charles R. Données de base pour la fumure des grandes cultures et des herbages (DBF-GCH). Revue suisse d'agriculture. 2009;**41**:1-98

[61] Sena MM, Frighetto RTS, Valarini PJ, Tokeshi H, Poppi RJ. Discrimination of management effects on soil parameters by using principal component analysis: A multivariate analysis case study. Soil and Tillage Research. 2002;**67**:171-181

[62] Shukla MK, Lal R, Ebinger M. Determining soil quality indicators by factor analysis. Soil and Tillage Research. 2006;**87**:194-204

Chapter 3

Composting

Niladri Paul, Utpal Giri and Gourab Roy

Abstract

Decomposition followed by stabilization of organic substances by biological actions has been taking place in nature from the very beginning of life appeared on our planet. Anthropogenic control and utilization of the process for sanitary disposal and reclamation of organic waste material have been termed composting and the final product is named compost. Microbial community leads the processes of both aerobic and anaerobic composting and converts wastes to a stable form of nutrients. The C/N ratio is the most important factor for decomposition, especially aerobic decomposition. Microorganisms respire two-third of carbon as CO_2, and one-third combines with nitrogen in living cell, and huge amount of heat energy is released as end product of aerobic decomposition as compared to anaerobic process. In agricultural world, utilization of human and animal wastes has great importance. Extensive studies on composting were initiated in India. Different composting methods like pit method, heap method, ADCO method, vermicomposting, etc. presently exist in the world. Humus is the end product of composting, and different organic wastes contain macro, micro, and trace elements, which reflect valuable properties for growing vegetation and to the soil itself.

Keywords: composting, anthropogenic, microorganism, vermicomposting, humus, elements

1. Introduction

Composting is a very old art, and some of its basic principles have been appreciated and used in practice for centuries. In recent years, however, rapid progress has been made in scientific studies of the underlying biological and chemical processes involved in composting. These studies have served to clarify several factors which can act to produce finished compost which is both valuable to agriculture and relatively safe from the viewpoint of public health [1].

There is an important relationship between sanitation and agriculture in all parts of the world. In agricultural areas, the utilization of human and animal wastes is of great importance from both the public health and the agricultural points of view. This is because of (a) the ever-increasing difficulties in disposing of great accumulations of wastes, (b) the ever-increasing threat to soil fertility, and (c) the intensive ever-increasing waste demand for agricultural lands to produce more food.

Sir Albert Howard and his associates [1] first suggested modern composting through his book *An Agricultural Testament* (1940) [2]. They studied in India, which was which was carried forward by Acharya and Subrahmanyan [3], further has been investigated extensively by Scott [4] and van Vuren by Gotaas [5] and his associates—McGauhey, Golueke, and Card—at the University of California [6], and by many others in different parts of the world.

 IntechOpen

2. Decomposition

Decomposition or stabilization of organic matter by biological action is the most valuable portion of life cycle on our planet. In recent times, man has attempted to control and directly utilize the process for sanitary disposal and reclamation of organic waste material, and this process has been termed "composting," and the final product of composting has been called "compost" [1].

Generally speaking there are two processes: (a) aerobic decomposition and stabilization and (b) anaerobic fermentation. In these processes, microbial community feed upon organic materials such as vegetable matter, animal manure, night soil, and other organic refuse and convert the wastes to a more stable form.

2.1 Aerobic decomposition

When organic material is decomposed in the presence of oxygen, the process is called aerobic. In aerobic stabilization, living organisms, which utilize oxygen, feed upon the organic matter and develop cell protoplasm from the nitrogen. From the nitrogen, phosphorus carbon and other required nutrients. Much of the carbon serves as a source of energy for the organisms and is burned up and respired as carbon dioxide. Since carbon serves both as a source of energy and as an element in the cell protoplasm, much more carbon than nitrogen is needed. Generally about two-thirds of the carbon is required as carbon dioxide (CO_2), while the other third is combined with nitrogen in the living cells. If the excess of carbon over nitrogen in organic materials being decomposed is too great, biological activity diminishes, and several cycles of organisms may be required to burn up most of the carbon. When some of the organisms die, their stored nitrogen and carbon become available to other organisms. The utilization of nitrogen from the dead cells by other organisms to form new cell material requires the burning of excess carbon to CO_2. Thus, the amount of carbon is required, and the limited amount of nitrogen is recycled. Finally, when the ratio of available carbon to available nitrogen is sufficiently low, nitrogen is released as ammonia. Under favorable conditions, some ammonia may be oxidized to nitrate. Phosphorus, potash, and various micronutrients are also essential for biological growth. These are normally present in more than adequate amounts in compostable materials and present no problem; hence, a discussion of their metabolism by the biological cells will not be included [1]. The cycle of nitrogen and carbon in aerobic decomposition is structured in **Figure 1**.

2.2 Anaerobic decomposition

Putrefactive breakdown of organic material takes place anaerobically. Anaerobic living organisms in metabolizing nutrients break down the organic compounds by a process of reduction. As in aerobic process, the organisms use nitrogen, phosphorus, and other nutrients in developing cell protoplasm but reduce organic nitrogen to organic acids and ammonia. Carbon from organic compounds which is not utilized in the cell protein is liberated mainly in the reduced form of methane (CH_4). A small portion of carbon may be respired as CO_2 [1].

This process takes place in nature as in the decomposition of organic muds at the bottom of marshes and in buried organic material to which oxygen does not have access. The marsh gas which rises is largely CH_4. Intensive reduction of organic matter by putrefaction is usually accompanied by disagreeable odors of hydrogen sulfide and of reduced organic compounds which contain sulfur, such as mercaptans [1].

Since anaerobic destruction of organic matter is a reduction process, the final product, humus, is a subject to some aerobic oxidation when put on the soil. This oxidation is minor, takes place rapidly, and is of no consequences in the utilization of the material on the soil [1]. The cycle of nitrogen and carbon in anaerobic decomposition is structured in **Figure 2**.

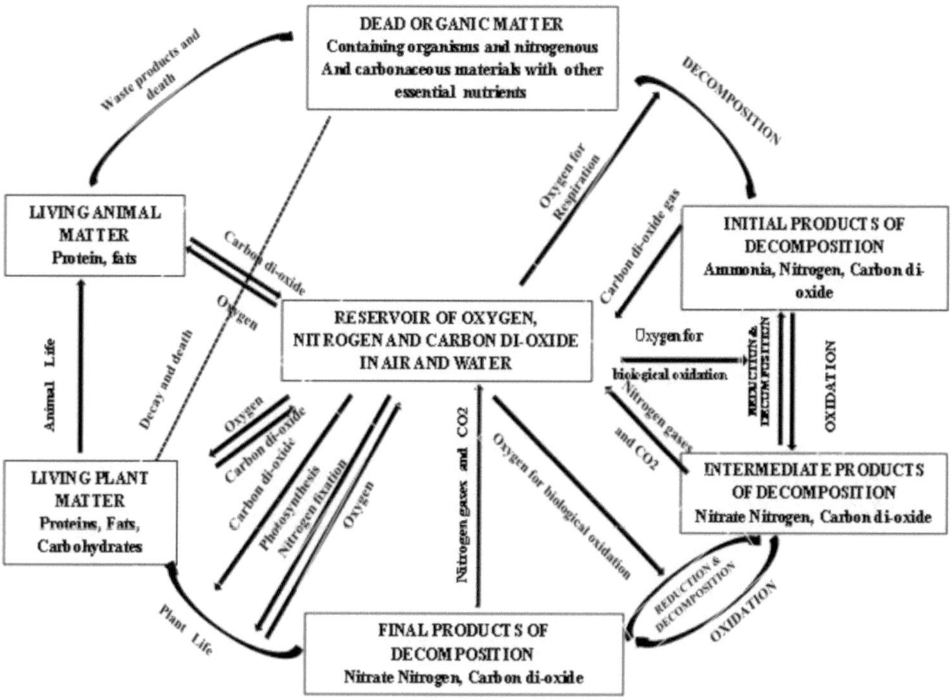

Figure 1.
Cycle of nitrogen and carbon in aerobic decomposition [7].

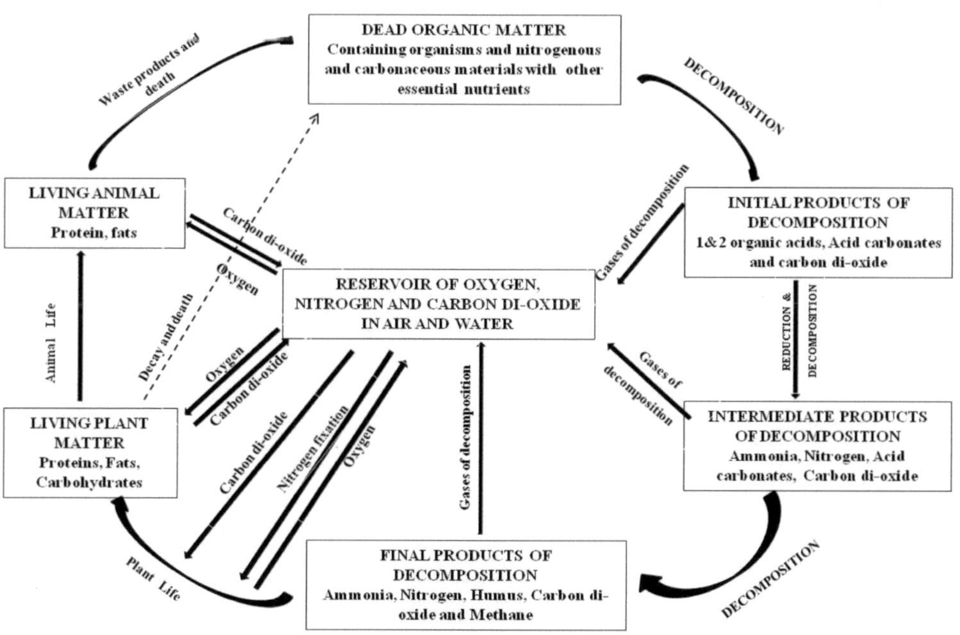

Figure 2.
Cycle of nitrogen and carbon in anaerobic decomposition [7].

3. Raw material

The quantity, characteristics, and composition of wastes available for composting vary widely with season and different localities. The multiplicity and complexity of the factors affecting the quality and quantity of compostable refuse prohibit the use of any formula or rule-of-thumb method for determining the amount of waste material to be expected at any given place [1]. Either a study of specific place or the use of information obtained from studies of places with very similar characteristics is necessary for estimating the quality and quantity of refuse for a given population. These are basic information, useful in supplementing local data in analyzing a particular composting operation.

In a particular agricultural village, following basic quantity and quality data will be useful for studying a compost operation.

3.1 Human feces without urine

Approximate quantity: 135–270 g per capita per day moist weight and 35–70 g per capita per day dry weight

Approximate composition: Moisture, 66–80%; organic matter (dry basis), 88–97%; nitrogen, 5.0–7.0%; phosphate (P_2O_5), 3.0–5.4%; potash (K_2O), 1.0–2.5%; carbon, 40–55%; calcium oxide, 4–5%; C/N ratio, 5–10 [8]

3.2 Human urine

Approximate quantity: 1.0–1.3 liters per capita per day and 50–70 g per capita per day

Approximate composition: Moisture, 93–96%; organic matter (dry basis), 65–85%; nitrogen, 15–19%; phosphate (P_2O_5), 2.5–5.0%; potash (K_2O), 3.0–4.5%; carbon, 11–17%; calcium oxide, 4.5–6% [8]

3.3 Animal manure

The quantity of animal manure varies widely with different conditions of feeding and stabling. Van Slyke [9] gave the information shown in **Table 1** on animal excrement production.

The stable manure is approximately composed with three main components: (a) bedding or vegetable matter litter, (b) solid excreta, and (c) urine. The characteristics and relative concentration of these components vary widely, depending on the type of animal, the stable feeding and handling, and the use to which the animal is put. Straw and plant residues used for bedding usually contain large amounts of carbon, particularly in the form of cellulose and small amounts of nitrogen and minerals. Considerable amount of protein is present in the solid excreta and provide balance nutrient material for the growth of microorganisms [1]. **Table 2** [10] reflects the chemical constituents in fresh manure from different animals, and **Table 3** [11] shows the chemical nature of different types of manure.

3.4 Refuse (garbage, rubbish, other litter)

The most available quantities of garbage, organic rubbish, and dead vegetables are used for animal feed. There is also little waste paper, rags, etc. in the refuse. Ash, particularly in cold climate, street sweeping, and trash constitute a major portion of waste. In warm areas with high rainfall, much waste vegetation finds its way into the refuse. However, in many villages the amount of such refuse is sufficient in

Animal	Tonnes per year per 454 kg live weight	Nitrogen (kg per year per 454 kg live weight)		
		Liquid	Solid	Total
Horse	9.00	2.5	3.8	6.3
Cow	13.5	2.2	2.2	4.4
Pig	15.3	1.8	1.6	3.4
Sheep	6.3	4.5	4.9	9.4
Poultry	4.3	—	9.1	9.1

Table 1.
Quantities of animal excrement [9].

Chemical constituents	Sheep manure	Horse manure	Cow manure
Ether-soluble substances	2.8	1.9	2.8
Cold-water-soluble organic matter	19.2	3.2	5.0
Hot-water-soluble organic matter	5.7	2.4	5.3
Hemicelluloses	18.5	23.5	18.6
Cellulose	18.7	27.5	25.2
Lignin	20.7	14.2	20.2
Total protein	25.5	6.8	14.9
Ash	17.2	9.1	13.0

Table 2.
Chemical composition of fresh manure from various animals (on the basis of dry, litter-free material) (in %) [10].

Manure	Moisture (%)	Composition of dry matter		
		Nitrogen (%)	Phosphate (%)	Potash (%)
Cattle	80	1.67	1.11	0.56
Horse	75	2.29	1.25	1.38
Sheep	68	3.75	1.87	1.25
Pig	82	3.75	3.13	2.50
Hen	56	6.27	5.92	3.27
Pigeon	52	5.68	5.74	3.23

Table 3.
Chemical nature of different types of manure [11].

quantity to provide a satisfactory compostable mass when mixed with night soil and animal manure. The approximate quantity of garbage in village is usually 220–340 g per capita per day with the following composition: moisture content, 10–60%; organic content (dry basis), 25–35%; nitrogen, 0.4–0.8%; phosphate, 0.2–0.5%; potash, 0.8–1.5%; carbon, 12–17%; and calcium oxide, 4.0–7.5% [1].

3.5 Slaughterhouse wastes

The amount of these wastes is extremely variable, depending upon the extent of processing. In small house with no by-product processing, the compostable wastes will be as much as 22–36 kg (dry basis) per ton of meat processed, while in large plant with by-product processing, the compostable wastes will be 11–18 kg (dry basis) per ton. The composition of slaughterhouse waste varies with the extent of utilization of wastes for the manufacture of by-products. Most rural slaughter-houses have primitive recovery processes, and the wastes consist of blood, unsalable meat, intestines, offal, paunch manure, hoofs, etc. and have the following average composition: moisture content, 75–80%; organic matter (dry basis), 80–95%; nitrogen, 8–11%; phosphate, 3.0–3.5%; potash, 2.0–2.5%; carbon, 14–17%, and calcium oxide, 3.0–3.5% [1].

4. Cities and urban centers

Compostable urban wastes probably vary as to quantity and composition almost as much as do rural wastes. Some basic data pertaining to cities with water-carried sewage collection and regularly operated garbage and refuse collection systems that can supplement local information in analyzing municipal composting operations will be shown. Sewage sludge, either fresh or digested, can be composted with garbage and other refuse with sufficient moisture so that the mass will compost aerobically. The quantities and composition of sewage solids and of the sludge are shown in **Table 4**.

In industrial areas, the waste composition and quantity vary with the type of industry. Domestic and food establishment waste garbage quantity depends on climate, food-preservation facilities, type of food used, and utilization of garbage for stock food and the economic status of the community. Domestic wastes vary from 90 to 400 g per day per capita with 60–85% moisture and 65–85% organic matter on dry weight basis. On the other hand, quantities of nonconsumable and non-compostable rubbish such as cans, bottles, china, and metal vary from 45 to 500 g per capita per day [5].

5. Different methods of composting

5.1 Indore method

During the early days of organic gardening/farming, this method was the only systematic way to mature compost. This method developed at the Institute of Plant Industry, Indore, India, between 1924 and 1931, was designed and described by Sir Albert Howard, known as the father of modern organic farming, in his dissertation on organic agriculture *An Agricultural Testament* (1940). In this method, animal dung is used as the catalytic agent along with different types of organic wastes available on the farm.

The steps followed for preparation of compost by Indore method are given below:

i. A compost heap of suitable size say 3 m × 1.5 m × 1 m (length × width × depth) is prepared. The selected site should be near the cattle shed and water source and at an elevated site so that no rainwater floods into the pit during rainy season.

ii. Organic wastes of different sources available on a farm are accumulated near the trench and mixed thoroughly. Hard woody materials (not exceeding 10% of the total plant residues) are crushed before being piled. Green materials,

	Quantity of solid (dry basis) g/head/day	Liquid sludge (% solid)	Drying bed cake (% solid)	Vacuum filter cake (% solid)	Composition on dry basis (%)				
					Organic	Mineral	Nitrogen	Phosphate	Potash
1. Fresh domestic sewage	81.6–99.7	0.04–0.15	—	—	60–85	15–40	5.0–10.0	2.5–4.5	3.0–4.5
2. Imhoff tank	22.7–36.3	8.0–12.0	35–50	—	30–45	55–70	2.0–3.0	1.2–3.5	0.1–0.5
3. Primary, fresh	45.4–63.5	2.5–5.0	28–45	22–34	60–80	20–35	1.5–4.0	0.8–4.0	0.1–0.5
4. Primary digested	27.2–40.8	5.0–12.0	35–50	26–34	35–60	40–65	1.0–3.5	1.2–4.0	0.1–0.5
5. Primary and trickling filter, humus fresh	59.0–77.1	3.5–6.5	26–40	23–34	50–75	25–50	2.0–4.5	0.8–3.6	0.1–0.5
6. Primary and trickling filter, humus digested	36.3–50.0	5.0–12.0	35–50	25–35	35–60	40–65	1.0–3.5	1.0–3.8	0.1–0.5
7. Primary and activated sludge, fresh	72.6–90.7	3.0–6.0	26–40	20–24	50–80	20–50	2.3–5.2	1.2–4.0	0.2–0.6
8. Primary and activated sludge, digested	45.4–59.0	45.8.5	28–50	22–26	35–55	45–65	2.0–4.8	1.3–4.0	0.2–0.6
9. Primary sludge, digested, and fresh activated sludge	54.4–72.6	2.5–4.5	28–45	20–24	40–60	40–60	2.2–3.0	1.3–4.0	0.3–0.8

Table 4.
Approximate quantity and composition of sewage and sewage sludge [5].

which are soft and succulent, are allowed to wilt for 2 to 3 days in order to remove excess moisture before stacking; these tend to pack closely when stacked in the fresh state. The mixture of different kinds of organic materials/residues ensures a more efficient decomposition [12].

iii. The compost heap is built in layers. First a layer of refuse/organic wastes like weeds, crop residue, grass clippings, or leaves of about 15–20 cm (6–8 inch) thick is spread at the base of the heap. Next a 5 cm (2 inch) layer of cattle dung slurry and water is added onto the refuse. A third layer of the same size of the first is then spread followed by a layer of slurry of cattle dung and water. This layering sequence is continued till the heap is raised to a height of 50–100 cm above the ground level. The top is then covered with a thin layer of soil, and the heap is kept moist.

iv. The filling of heap is completed within 6–7 days to fill the three-fourth length of the trench, leaving 1/4th length empty to facilitate subsequent turnings.

v. Water is sprayed on regular basis so as to keep the moisture content to about 60–80%.

vi. Turning is done three times, at 15, 30, and 60 days after compost filling in order to allow air to penetrate so that the heap will heat up properly. At each turning the whole mass is mixed thoroughly. This can be done manually or mechanically.

The main advantage of this method is that the finished compost is ready within 4–5 months for application to the soil. The composed prepared by this method contains, on an average, 0.8% N, 0.3–0.5% P_2O_5, and 1.0–1.5% K_2O. Periodic turning of composting mass helps the process to remain aerobic throughout the decomposition and facilitate faster decomposition by bringing the substrates which are undecomposed or partially decomposed with the microorganisms and air. As it requires extra labor, the cost of preparation of compost is more. Heat is generated during the decomposition process inside the compost pit which helps in destroying most of the pathogens and weed seeds. When sufficient nitrogenous material is not available, a green manure or leguminous crop like sunnhemp *(Crotalaria juncea)* may be grown on the fermenting heap after the first turning. The green matter is then turned in at the second mixing [12].

5.2 Bangalore method

This method is an anaerobic process, developed at the Indian Institute of Science, Bangalore, by the late Dr. C.N. Acharyain in 1939. It is recommended where night soil and refuse are used for preparing the compost. This method overcomes many of the disadvantages of the Indore method, such as the problem of heap protection from adverse weather, nutrient losses from intensive rains and strong sun, frequent turning requirements, and fly nuisance [12]. The method is suitable for areas with scanty rainfall. The compost is done in the trenches of 9.1 m × 1.8 m × 0.9 m (=302′ × 62′ × 32′) or in the pits of 6.1 m × 1.8 m × 0.9 m (= 202′ × 62′ × 32′). This method saves on labor cost because there is no need of turning and regular sprinkling of water but takes much longer time to finish [12].

This method includes the following steps:

i. As like Indore method, the mixed farm residues are spread at the bottom of a trench or pit of a convenient size, similar to that of Indore method. Generally, trenches or pits about 1 m deep are dug 1 m in breadth, and the length of the trenches can vary according to the availability of land and the type of material to be composted. The trenches should preferably have sloping walls and floor to prevent water logging.

ii. Organic residues and night soil are put in alternate layers. The trench or pit is filled layer-wise till the raw material reaches about 50 cm above the surface. Here 100% space of pit is used.

iii. The pit is covered with 15–20-cm thick layer of refuse and then plastered with a 2–5 cm layer of a mixture of mud and cattle dung. Plastering of pit prevents the loss of moisture and fly nuisance. This method effectively controls foul smell and kills pathogenic organisms.

iv. The materials are allowed to remain in the pit without turning and watering. During this period the material settles down due to reduction in the volume of biomass. Under such conditions, decomposition is largely anaerobic and high temperatures do not develop. The C/N ratio of the finished product drops to a value below 20:1 with no odor, indicating that the compost is ready to use.

v. The material undergoes anaerobic decomposition at a very slow rate, and it takes about 6–8 months to obtain the finished product.

vi. The recovery of the finished product is greater than aerobic composting.

vii. Labor requirements are less than for the Indore method as turning of material is not done; labor is needed only for digging and filling the pits.

Organic nitrogenous compounds gradually become soluble, and the carbonaceous matter breaks down into CO_2 and H_2O. The loss of ammonia is negligible because in high concentrations of CO_2, forming ammonium carbonate is stable. The anaerobic process is particularly suited for use by gardeners in or near cities and towns. The well-decomposed compost contains 0.8–1.0% N. A uniform high temperature is not assured in the biomass. Problems of odor and fly breeding need to be taken care of. After 8–9 months, all the material decomposes, and the compost becomes ready for application.

5.3 NADEP composting

This method of composting was developed by Sri Narayan Deorao Pandharipande. He was an old Gandhian worker, popularly known as "Nadep Kaka" from Maharashtra. He worked for 25 years at the Dr. Kumarappa Gowardhan Kendra at Pusa to perfect his composting technique [2]. This process facilitates aerobic decomposition of organic matter. This method takes care of all the disadvantages of heaping of farm residues and cattle shed wastes in the open condition. This method envisages a lot of composting through minimum use of cattle dung. It requires composting materials like dung, farm residues, soil, waste products of agriculture, etc. Decomposition process follows the "aerobic" route and requires about 3–4 months for obtaining the finished product.

This method includes the following steps:

i. A brick structure measuring 9′ × 6′ × 3′ with perforated holes in all the side walls is prepared to ensure adequate supply of air during composting. It is carried out in specially constructed tanks with walls built like "honeycombs" through which water is sprayed to prevent the compost from becoming dry. This aboveground-perforated structure facilitates passage of air for aerobic decomposition. The floor of the tank is laid with bricks and covered above with a thatched roof. This prevents loss of nutrients by seepage or evaporation, and the contents are not exposed to sunshine and rain [12].

ii. The brick tank is plastered with cattle dung slurry to facilitate bacterial culture for decomposition of biodegradable wastes. The brick tank is then filled layer-wise first with a thick layer (10–15 cm) of chopped fine stick of semihard wood which helps in providing aeration, followed by a same layer of farm wastes or dry and green biomass or any other biodegradable material to be composted.

iii. Prepared slurry of mixing cattle dung (5–10 kg) with water (100 liters) is then sprinkled thoroughly on the biodegradable mass in order to facilitate bacterial culture for faster decomposition. On it a layer of soil is maintained in order to compress the volume of the wastes. Addition of soil also facilities retention of moisture, provides microorganisms, acts as buffer, and controls pH of the compost during decomposition. The nutrients produced in the manure are absorbed by the soil layers, thus preventing nutrient loss.

iv. The whole tank is thus filled completely with about 10–12 layers in the same sequence having 1–3 sub-layers in each layer. After 2–4 weeks, the volume of the composting mass is reduced to almost two-third of the original. At this stage, additional layers of composting mass are formed over it keeping the same sequential set up, already said. Finally, the whole biomass is plastered and sealed with slurry of cattle dung and mud. In this condition, the tank is allowed to decompose the biodegradable wastes for further 3 months. Water is added on regular basis to maintain the moisture content between 60 and 75% throughout the composting period.

v. It is advisable to sprinkle microbial cultures like *Trichoderma*, *Azotobacter*, and *Rhizobium* and phosphate-solubilizing microorganisms in each layer to enhance the equivalent speed of composting process at each corner of the compost.

vi. Compost becomes ready for use within 110–120 days after composting. So one tank can be used three times annually.

vii. The prepared compost can be stored for future use, preferably in a thatched shed after air-drying and maintaining it at about 20% moisture level by sprinkling water whenever needed. Also storage at gunny bag in shade areas is also preferable. By following this procedure, the composed could be preserved for about 6–8 months.

viii. Requirement of higher labor and inconvenience faced in filling during rainy season are the two difficulties experienced by the farmers in adopting NADEP method of composting.

5.4 Municipal solid waste composting (MSW)

MSW composting or mechanical composting is followed in big cities, where huge quantities of garbage are generated. The metropolitan cities like Mumbai, Kolkata, Delhi, and Chennai generate about 2000–6000 tonnes garbage per day, posing gigantic disposal problems. Mechanical composting plants with capacity of 500–1000 t day-1 of city garbage could be conveniently installed in big cities and 200 t day-1 plants in the small towns in India. The adoption of accelerated fermentation treatment enables 70% of the refuge to be available as refined compost in the form of a dry, black free-following material, easy to transport and handle. Such refined mechanical compost contains generally equivalent amount of mineral matter and organic matter with half of organic carbon. The composition of the compost is variable and at par with the raw materials used. On an average, it may contain 0.7% N, 0.5% P_2O_5, and 0.4% K_2O and a C/N ratio of 15–17. Mechanical composting has several advantages such as (i) environmental sanitation to minimize pollution, (ii) recycling of discarded wastes into a value-added product, and (iii) production of compost within a short period [12].

5.5 Enriched compost

In general, the bulky organic manures like FYM contains around 0.5–1.0% N, 0.2–0.5% P_2O_5, and 0.5–1.0% K_2O. The cost of preparation, storage, transport, and application of FYM or compost to soils is high. The demerits of bulky manures can be overcome through the preparation of enriched compost by adding nitrogen, phosphorus, potassium, and micronutrients either alone or in combination [13].

Enriched composts have the following advantages:

i. Enriched compost is more concentrated than compost; it reduces the bulk to be handled per unit of nutrient.

ii. It may increase nutrient use efficiency of added fertilizer and maintain soil organic carbon.

iii. It prevents nutrient losses due to microbial immobilization of nutrients during decomposition of organic residues and due to adsorption of cations on account of high exchange capacity of organic matter.

iv. Lesser problems in handling, storage, and transportation.

v. Offers a potential avenue for the efficient utilization of low-grade materials such as rock phosphate and waste mica (a K-bearing mineral)

Enrichment of compost can be done in two ways, namely, (i) physical addition of fertilizer materials during composting and (ii) addition of fertilizer materials with ready compost by mixing. Incorporation of fertilizers during composting leads to immobilization of fertilizers into microbial body and insertion into molecules of humic substances formed during decomposition. A substantial part of added inorganic nutrients may also be adsorbed on to exchange sites or chelated by humic substances. On the other hand, physical mixing of fertilizers with finished product of compost reflects adsorption and chelation of fertilizer elements by humic substances, which are already present in the decomposed product [13].

5.5.1 Enrichment with nitrogen

Microbial mineralization and immobilization depend on the C/N ratio of the compost. The wide C/N ratio (>30:1) plant materials require addition of mineral N to narrow down the C/N ratio for rapid decomposition including mineralization during composting. During the preparation of compost from wide C/N ratio substrates, such as straws/stubbles, incorporation of fertilizer N like ammonium sulfate or urea at 0.5–1.0% of raw materials hastens the decomposition process. Addition of nitrogenous fertilizer serves as starter. Enrichment of N during composting with inorganic N can be done up to 1.8–2.5% but cannot be improved beyond 2.5% N, because of the associated losses of N includes the production of free NH_3.

In case of ready compost, it is recommended that compost with a C/N ratio of about 20:1 should be treated with fertilizer nitrogen so as to bring the C/N ratio to <10:1 and N content >2.5%. Thus, by spraying a solution of urea on finished product of compost followed by physical blending, the N content can be increased up to 5–7%. As most of the added inorganic N remains in the fertilizer from without much of chemical or biological reaction with the manure, it is tough to understand the utility of using fertilizers to raise N content of the finished product above 5–7%.

5.5.2 Enrichment with phosphorus

Phosphorus-enriched compost can be prepared by adding 5% superphosphate, dicalcium phosphate (DCP), and rock phosphate at the time of filling of the compost pits. Due to enrichment with soluble phosphate in compost, a small amount of immobilized soluble P into microbial body may be expected. But with most plant material containing sufficient P to satisfy microbial demands during decomposition, assimilation of P from external sources is seldom needed. Addition of insoluble sources of P like low-grade rock phosphate to enrich compost is a more rational and practical approach, since solubilization of sparingly soluble P occurs during composting. Besides phosphorus, it is a source of calcium and micronutrients. Early work showed that by adding rock phosphate to farm composting materials to a thickness of about 5 mm per layer, nearly 50–70% of sparingly soluble P could be converted to soluble from which is readily available to plants. Addition of soluble fertilizer-P to finished compost provides a better scope for increasing the efficiency of fertilizer-P as well as organic-P. Thorough mixing of fertilizers with compost may reduce P-fixation. The mineralization of organic-P may also be accelerated due to increased solubility of organic-P in the presence of fertilizers. Amalgamation of compost with single superphosphate (SSP) could raise phosphorus content of the enriched compost up to 5% P_2O_5 [13].

5.5.3 Enrichment with potassium

To enrich the compost, potassium-bearing minerals like feldspars and mica can be added during composting. The availability of potassium can be improved due to the production of organic acids such as citric, tartaric, acetic acid, etc. Potassium can also be added to compost by incorporating plant materials, which contain appreciable amounts of potassium, viz., water hyacinth and banana skin, are rich source of potassium. Dry potato vines also contain about 1% potassium which can be incorporated to improve the K content in the compost [13].

5.5.4 Enrichment with bioinoculants

Addition of nitrogen-fixing bacteria and/or phosphate- and potassium-solubilizing microorganisms is one of the possible means of improving nutrient content of the final product of compost. Inoculation of *Azotobacter*, *Azospirillum*, *Clostridium*, etc. to the compost heap enhances N content by fixing atmospheric N_2. Phosphate-solubilizing bacteria such as *Bacillus polymyxa*, *Pseudomonas striata*, and fungi such as *Aspergillus awamori* can be introduced into the composting mass along with rock phosphate. These microorganisms help in solubilizing sparingly soluble inorganic phosphates due to the production of organic acids such as citric, tartaric, gluconic acid, etc. and thereby increasing the available P, both water-soluble and citrate soluble P, content of compost. Some cellulolytic and lignolytic microorganisms such as *Trichoderma viride*, *Trichurus spiralis*, *Paecilomyces fusisporus, and Phanerochaete chrysosporium* are used as compost accelerator to hasten the process of composting [13].

5.6 ADCO compost

This process was introduced in England in 1921. Hutchinson and Richards [14] developed an ADCO powder, used as a starter at 7.0 kg per 100 kg dry waste product. Fowler assured that this powder is prepared with various substances like ammonium phosphate, cyanamide, and urea. On the other hand, Collision and Conn prepared another powder of 27 kg ammonium sulfate, 13.5 kg superphosphate, 11.250 kg murate of potash, and 22.5 kg ground limestone and added to 1 ton dry matter for producing manure. This produced manure has characteristic resemblance with manure produced using ADCO powder. For ADCO process a plane place measuring 450 cm long and 180 cm breadth is required. First, a layer of refuse about 30 cm thick is spread at the bottom of the pit, and over this a calculated amount of ADCO powder, i.e., 7 kg per 100 kg refuse, is sprayed. Six-time addition of refuse in that pit means 1 ton refuse, and every time ADCO powder is added. The heap height should be within 180 cm, i.e., 6 feet. After completion of heap, time to time watering is done. Through aerobic composting, the manure becomes ready within 4–5 month.

Advantages: Very suitable method for making compost. Within 4–5 months proper decomposition makes good organic manure.

Disadvantages: Regular turning is required for aeration and watering for proper decomposition. It increases labor charges and cost of production.

5.7 Vermicompost

Compost prepared using earthworms is called vermicompost. Earthworms consume all type of organic matter especially green matter, retain 5–10% for their growth, and excrete the mucus-coated undigested matter called vermicast. This undigested matter undergone physical and chemical breakdown by the activity of muscular gizzard present in the worms' intestine. It is a cost-effective, time saving, and efficient process of recycling nontoxic animal and agricultural and industrial wastes. Vermicast is rich in nutrients—N, P, K, Ca, Mg, vitamins, enzymes, and growth-promoting substances. In addition, the warms do the turning and no additional turning of the compost heap is required. The efficient species of earthworms are *Eisenia foetida*, *Pheretima elongata*, *Eudrilus eugeniae, and Perionyx excavatus* [13].

For preparation of a good quality of vermicompost, a number of steps are followed as mentioned below:

i. Selection of earthworm: The locally available earthworm native to a particular soil and efficient for fast composting may be used for vermicomposting.

ii. Size of pit: Any convenient dimension such as 2 m × 1 m × 1 m may be prepared. This can hold 20,000–40,000 worms giving one ton manure per cycle. The pit should be base concreted as termite proof and ant proof through water drain around it. A shade of 6–8 ft height is also required for cool and ambient climate for the worms.

iii. Preparation of vermibed: A thick layer of 15–20 cm of good loamy soil above a thin layer (5 cm) of broken bricks and sand should be made. This layer is prepared on concreted floor and made to inhabit the earthworms.

iv. Inoculation of earthworms: About 100 earthworms are introduced as an optimum inoculating density into a composite pit of about 2 m × 1 m × 1 m, provided with a vermibed.

v. Organic layering: It is done on the vermibed with fresh cattle dung of 5–10 cm. The compost pit is then layered to about 5 cm with dry crop residues. Carbon-rich solid and dead substrates like sawdust, paper, and straw are mixed with N-rich natural components such as sewage, sludge, and biogas slurry to obtain a near optimum C/N ratio. Mixing variety of substances produces good-quality compost which is rich in macro, micro, and even trace nutrients. Decomposition can be accelerated by chopping raw materials into small pieces. Moisture content of the pit is maintained at 50–60% of water holding capacity. Aeration can be maintained by mixing with fibrous N-rich materials. The temperature of the piles should be around 28–30°C. Wide gap between higher or lower temperatures reduces the activity of microflora and earthworms. The normal pH of the raw materials is preferable.

vi. Wet organic layering: It is done after 1 month with moist/green organic waste, which can be spread over it. This practice can be repeated every 3–4 days as per requirement. Mixing of wastes periodically without disturbing the vermibed ensures proper vermicomposting. Wet layering with organic waste can be repeated till the compost pit is nearly full.

vii. Harvesting of vermicompost: In order to facilitate the separation of worms from vermicompost, the moisture content in the compost is brought down by stopping the addition of water around 7–10 days before maturation that ensures drying of compost and migration of worms into the vermibed. This forces about 80% of the worms to the bottom of the bed. The remaining worms can be removed by hand. The mature compost, a black, fine loose, granular humus rich material, looks like CTC tea, is removed out from the pit, dried, and packed. The pleasant earthen smell is one of the good indications of mature compost. The vermicompost is then ready for application.

The nutrient content of vermicompost varies depending on the raw materials as well as different species of earthworms used. Thus, the final product is not a single standard product. The average nutrient content of vermicompost is 0.6–1.2% N, 0.13–0.22% P_2O_5, 0.4–0.7% K_2O, 0.4% CaO, and 0.15% MgO. On an average, it contains comparable N, P, and wide C/N ratio as in FYM but less K and micronutrients than FYM. On the whole, vermicompost cannot be described as being nutritionally superior to other organic manures. Yet the unique way in which it is produced, even in the field condition, time saving, and at low cost, makes it very attractive for practical application. Unique feature of vermicompost is its rapid process of composting which takes about 60–90 days depending on the environmental conditions. The excess worms that have been harvested from the pit can be used in the other pits, sold to other farmers for compost inoculation, and may be used as animal and poultry feed or fish food [13].

6. Green manure

Green manuring is the practice of enriching soil nutrient status by growing a crop and plowing in situ or turning it into the soil as undecomposed green plant materials for the purpose of improving soil health. These crops are known as green manure crops. They improve soil physical properties and supplies nutrients particularly N, if it is a legume crop. Green manuring can be of two types.

6.1 In situ green manuring

When the green manure crop is grown and buried in the same field, it is called in situ green manuring. Most important in situ green manuring crops are sunnhemp (*Crotalaria juncea*), dhaincha (*Sesbania aculeata*), cowpea (*Vigna sinensis*), berseem (*Trifolium alexandrinum*), and Lucerne (*Medicago sativa*) [15].

6.2 Green leaf manuring

These are the plants grown elsewhere, and green leaves and tender twigs are brought to the field for incorporation. This is labor consuming. Popular green leaf manuring plants are *Leucaena leucocephala* (Subabul), *Cassia tora*, *Sesbania speciosa*, *Pongamia pinnata* (Karanj), *Pongamia glabra*, and *Gliricidia maculata* [15].

In general, green manure crops should be a legume with good nodulation, i.e., N_2-fixing capacity, fast growing, having low water requirement, and short duration, i.e., 4–6 weeks with tender leafy habit permitting rapid decomposition. Incorporation of green manure crop should be done before or at flowering stage because these are easily decomposed at this stage after which these become fibrous and take more time for decomposition.

7. Concentrated organic manures

These manures contain higher percentages of major essential plant nutrients (N, P, and K) compared to bulky organic manures (FYM and compost). They are derived from raw materials of plant or animal origin, such as oilcakes, fish manure, dried blood, bone meal, etc. Oilcakes are the residues, left after oil is extracted from oil-bearing seeds. Generally, edible oilcakes are used for animal feed, while nonedible oilcakes are used as manures. Oilcakes contain higher amounts of N than P_2O_5 and K_2O; thus, these are commonly referred to as the organic nitrogenous fertilizers. Bone meal consists of calcium phosphate together with fats and proteins. These are good sources of lime, phosphate, and N. Bone meal is a slow-acting organic-P-fertilizer resembled with rock phosphate and suitable for acid soils. Fish manure is a quick-acting manure and suitable for all soils and crops. It is available as either dried fish or fish meal or powdered fish. However, its use is restricted mainly to coastal areas where it is available easily. Guano (dried excreta of sea birds) is

Product	N (%)	P_2O_5 (%)	K_2O (%)
Plant origin			
Edible oil cakes			
Safflower (decorticated)	7.9	2.2	1.9
Groundnut	7.3	1.5	1.3
Sesame	6.2	2.0	1.2
Rapeseed/mustard	5.2	1.8	1.2
Linseed	4.9	1.4	1.3
Nonedible oil cakes			
Neem	5.2	1.0	1.4
Castor	4.3	1.8	1.3
Karanj	3.9	0.9	1.2
Cottonseed (undecorticated)	3.9	1.8	1.6
Mahua	2.5	0.8	1.8
Animal origin			
Blood meal	10–12	1.0–2.0	0.6–0.8
Meat meal	10–11	2.0–2.5	0.7–1.0
Fish meal	5–8	3.0–6.0	0.3–1.5
Guano	7–8	11–14	2.0–3.0
Slaughterhouse waste	8–10	3.0	—
Bone meal (raw)	3.0	20.0 (8% citrate soluble P_2O_5)	—
Bone meal (steamed)	—	22.0 (16% citrate soluble P_2O_5)	—
Wool waste	4–7	—	1.0–5.0
Miscellaneous			
Press mud	1.0–1.5	4.0–5.0	2.0–7.0

Table 5.
Average nutrient content in concentrated organic manures [6].

another concentrated organic manure, containing substantial amount of nutrients, particularly N and P_2O_5, but it is not produced in India [13]. Average nutrient contents in various concentrated organic manures are placed in **Table 5**.

8. Sewage and sludge

Sewage refers to the liquid portion, and sludge refers to the solid portion of the waste which originates from the city sewerage system. Raw sewage consists mainly of water carrying suspended and dissolved black colored solid organic matter which may pollute water bodies (rivers). For that reason, it is treated by some means to reduce the organic matter load before it could be disposed off safely. During siphoning at sewage treatment plant, the sludge portion settles down and is separated from the liquid portion (sewage). The sewage can be used for irrigation purposes, while sludge can be used as manure as it contains large amount of plant nutrients. It has been estimated that available sewage of big cities in India could annually contribute around 1.2 Mt of N, 1.0 Mt of P_2O_5 and 0.8 Mt of K_2O. However, it contains excessive organic and N loading, and repeated application of untreated sewage water can result in soil sickness due to anaerobiosis and imbalance in C/N and C/P ratio and clogging of soil pores by colloidal matter and bacterial contamination of vegetables grown using them. Treated sewage water, after dilution (1:1) with good-quality water, can increase yield of crops. The main disadvantage of using sewage and sludge in agriculture is its heavy metals content, particularly Pb, Cd, Cr, and Ni depending on the source of industry from where the sewage and sludge originates. Thus, repeated application of sewage tends to increase the concentration of metals in soils and their availability to plants, which in turn could get into our food chain [16].

9. Distillery effluents (spent wash)

It is the by-product of manufacturing of ethyl alcohol from molasses. It contains considerable amounts of organic matter and plant nutrients especially K and S and appreciable amounts of N and P. This can be applied as irrigation water and as an amendment (for alkali soils). However, because of its high organic load, it may results biochemical oxygen demand (BOD) and chemical oxygen demand (COD) in water. For that reason, they are unsafe for direct application on to agricultural lands. Spent wash can, however, be safely applied to different crops after suitable dilution and has been reported to increase yield of several crops. Treatment of this effluent through biomethanation digesters reduces the organic matter load but still carries considerable organic and salt load, making its disposal a problem [15].

10. Conclusion

Composting is a natural phenomenon and pervasively relates with organic farming. Accelerating the quality and speed of compost is a scientific phenomenon and irrevocable for sustainable growth and development of agriculture without any toxic effect on environment and livelihood.

It is established that any single method or technique of composting cannot be recommended for all areas and conditions. Also area-wise economic, climatic, social, and other factors will dictate the best method for that area. The efficiency of composting technique also depends on the type and amount of substrate(s) and the rearing techniques. However, it is hoped that the described methods will aid economic improvement in many areas and help establishing sustainable agriculture for the betterment of future. In consideration of time and quality, vermicomposting seems to be the best technique for composting and much more economically viable for the sustainable growth and development of modern agriculture. Vermicomposting technique is also worm and site specific. After long-term scientific experiments, *Eisenia fetida* is considered as the world's most efficient species having the capacity to acquaint with wide environmental condition. The compost production capacity of this worm is higher than other species, and so this species is widely accepted for vermicomposting.

Author details

Niladri Paul[1*], Utpal Giri[2] and Gourab Roy[3]

1 Department of Soil Science and Agricultural Chemistry, College of Agriculture, West Tripura, India

2 Department of Agronomy, College of Agriculture, West Tripura, India

3 Department of Zoology, Maharaja Bir Bikram College, Tripura, India

*Address all correspondence to: nilupaul82@gmail.com

References

[1] Howard A. The waste products of agriculture: Their utilization as humus. Journal of Royal Society of Arts. 1933;**84**:11, 13, 15, 25, 34, 36, 37, 38

[2] Biswas DR, Ghosh AK. Manures, biofertilizers and fertilizers. In: Soil Science, an Introduction. Indian Society of Soil Science. New Delhi: National Agricultural Science Centre Complex. ISBN: 81-903797-7-1. p. 436

[3] Acharya CN. Preparation of Compost Manure from Town Wastes. Calcutta: Indian Council of Agricultural Research, Ministry of Agriculture; 1950. Bulletin No. 60

[4] Scott JC. Health and Agriculture in China. London: Faber; 1952. 280 p

[5] van Vuren JPJ. Soil Fertility and Sewage; 1949

[6] University of California, Sanitary Engineering Laboratory. Reclamation of Municipal Refuse by Composting. Burkeley, California (Technical Bulletin No. 9); 1953

[7] Imhoff K, Fair GM. Sewage treatment. In: Composting. New York: World Health Organization; 1956. pp. 14, 16

[8] Gotaas HB. Raw material: Quantity and composting. In: Composting. World Health Organization; 1956. pp. 35, 40

[9] Millar CE, Turk LM. Fundamentals of soil science. In: Composting. 2nd ed. New York: World Health Organization; 1956. p. 36

[10] Waksman SA. Soil microbiology. In: Composting. New York: World Health Organization; 1952, 1956. p 36

[11] Jenkins SH. Organic manures. In: Composting. Harpenden, England: World Health Organization; 1935, 1956. p. 37

[12] Narayanasamy G, Arora BR, Biswas DR, Khanna SS. Fertilizers, manures and biofertilizers. In: Goswami NN, Rattan RK, Dev G, Narayanasamy G, Das DK, Sanyal SK, et al., editors. Fundamentals of Soil Science. New Delhi: Indian Society of Soil Science; 2009. pp. 579-622

[13] Biswas DR, Ghosh AK. Manures, biofertilizers and fertilizers. In: Rattan RK, Katyal JC, Dwivadi BS, Sarkar AK, Bhattacharyya T, Kukal JCTSS, editors. Soil Science: An Introduction. New Delhi: Indian Society of Soil Science; 2009. pp. 424-461

[14] Das PC. Manures. In: Manures and Fertilizers. Ludhiyana, India: Kalyani Publishers. p. 40

[15] Palaniappan SP. Green manuring: Nutrient potentials and management (chapter 4). In: Tandon HLS, editor. Fertilizers, Organic Manures, Recyclable Wastes and Biofertilizers. New Delhi: FDCO; 1994, 1994. pp. 52-71

[16] Ignatieff V and Page HJ (1958). Efficient Use of Fertilizers. F.A.O. Agricultural Studies No 43 (second edition)

Chapter 4

Plant Growth Biostimulants from By-Products of Anaerobic Digestion of Organic Substances

Sharipa Jorobekova and Kamila Kydralieva

Abstract

The by-products of anaerobic fermentation of agricultural wastes as a result of methanogenic microorganism activity are various bioactive substances, including humic-like substances (HLS). The contents of HLS formed are changed during fermentation process. The degree of humification significantly ranges on different fermentation stages reflecting the biosynthetic activity of microbial consortium in the processes of maturing and transformation of humic compounds. Characteristics of HLS isolated on various fermentation stages and bioactivity assessment present much interest for future applications as plant growth biostimulants, organic-mineral fertilizers, and phytohormones.

Keywords: humic-like substances, anaerobic fermentation, biostimulants, phytohormones

1. Introduction

Anaerobic digestion of organic wastes including manure with other substrates such as energy crops, industrial wastes, or food industry wastes is a commonly-used method as it can transform organic matter into biogas [1, 2]. During waste anaerobic digestion, the degradation of organic substances is commonly divided into the following stages: hydrolysis, acidogenesis, acetogenesis, and methanogenesis [3–6]. High molecular weight (MW) compounds, such as lipids, polysaccharides, proteins, and nucleic acids, are degraded into soluble organic substances (e.g., amino acids and fatty acids) and then split further into volatile fatty acids, ammonia (NH_3), CO_2, H_2S, and other by-products. The higher organic acids and alcohols are further digested to form mainly acetic acid, CO_2, and H_2, which are used to produce methane by different methanogens [7]. Besides proteins, polysaccharides, lipids, and nucleic acids, humic-like substances (HLS) are also major organic constituents of liquid digestate [3], sludge [8, 9], and their contents were reported to reach 26–28% of sludge organic matter [10, 11].

There are two major theories regarding HSs formation. Firstly, in the lignin theory, HSs are synthesized from precursors originating from lignin, meaning that lignin is the raw material and skeleton of HS precursors [12]. According to Kulikowska [13], the partial degradation of lignin can form phenolic and quinone moieties that can serve as HS precursors. Secondly, in the polyphenol theory, HSs are the condensation products of many small molecules, such as polysaccharides and proteins [12].

When compared with commercial humic substances, digester and sludge humic substances contain a wider variety of organic substances, more lipids, more nitrogen, and a lesser degree of oxidation. However, there are no significant differences in the effects of the two types of humic substances on plant growth [14]. Consequently, humic acids and fulvic acids could be extracted from digested sludge as a new source of organic liquid fertilizer, i.e., biostimulants. According to the definition by Jardin [15], plant biostimulant is any substance or microorganism applied to plants with the aim to enhance nutrition efficiency, abiotic stress tolerance, and/or crop quality traits, regardless of its nutrient content. Application of humic substances—soluble humic and fulvic acid fractions—shows inconsistent, yet globally positive, results on plant growth. A recent random effect meta-analysis of HS applied to plants [16] concluded on an overall dry weight increase of 22 ± 4% for shoots and of 21 ± 6% for roots. The variabilities in effects of HS are due to the source of the HS, the environmental conditions, the receiving plant, and the dose and manner of HS application [16].

2. Effects of anaerobic digestion on humic-like substance values

2.1 Extraction of humic-like substances

In order to extract humic acids and fulvic acids, some disintegration methods are applied to disrupt flocs and cells and release inner organic substances. The disintegration methods include mechanical, thermal, chemical, ultrasonic, and biological treatments [17]. Some methods can also be combined to disintegrate sludge [18–23]. As one of these methods, alkaline treatment has the advantages of simple devices, easy operation, and high efficiency. Alkaline sludge pretreatment was reported to enhance the dissolution of organic substances [24, 25] and also make humic substances be released from sludge particles [26]. Whether for sludge disintegration or for the extraction of humic substances, sodium hydroxide (NaOH) was more efficient than calcium hydroxide [$Ca(OH)_2$] [26]. After humic substances are transferred from waste solid phase into liquid phase by alkaline treatment, the dissolved humic substances in the supernatant can be recovered by ultrafiltration separation [2].

2.2 Evolution of humic-like substances during anaerobic sludge digestion

During waste anaerobic digestion, the humic substances evolved compared with that of other organic substances, i.e., proteins, polysaccharides, lipids, and nucleic acids. According to their solubilization in acidic or alkaline solution, HS can be divided into humic acids (HAs), fulvic acids (FAs), and humin (HU). HSs are generally recognized as being nondegradable or hard degradable during wastewater treatment processes, and the removal of HS from wastewater is attributed to biosorption of activated sludge instead of biodegradation [27]. However, HS may be generated by microbial activities during waste storage and treatment [28]. Sludge was found to be enriched in oxygen functional groups and aromatic rings during the course of storage and composting [29–31], but the humification process is not complete due to the typically low free radical concentrations and the unstable C/N ratio [32]. On the other hand, HSs were also found to be utilized by microorganisms as a supplementary source of nutrients [33]. HAs increased at first and then decreased in landfill composed of municipal refuse and sludge because both humification and mineralization processes took place [33, 34]. Most researches on HS evolution have been confined to sludge composting, landfilling, and storage. Few studies have investigated that HS might be dynamically involved in carbon and electron flow in anaerobic environments [35, 36]. This electron transfer would yield energy to support growth and stimulate the mineralization of organic compounds under anaerobic conditions [37, 38]. Bartoszek et al. [30] found that HAs became enriched in oxygen functional groups and aromatic rings in a digestion chamber.

During sludge anaerobic digestion, 16.3% of HAs and 27.0% of FAs were degraded, but the degradation rate was relatively low compared with that of other organic substances in sludge. Besides the mineralization of sludge HS, humification processes also took place. The HS extracted from the digested sludge have more oxygen functional groups, more aromatic structures, and larger molecular sizes compared with the HS extracted from the raw sludge. However, the degree of humification was low, and mineralization was still the main process that occurred during sludge anaerobic digestion [2].

2.3 Dynamics of physico-chemical parameters: case study

Fermentation process is carried out at the disposable or periodic loading of the bioreactor by common raw materials of fermentation such as animal manure, sewage sludge, food wastes, and green wastes (the optimal C/N ratio is 20–25). Animal manure and sewage sludge are characterized by a high moisture content, a low C/N ratio, and a low porosity [39, 40]. Green wastes have a low moisture content as well as a high C/N ratio, porosity, and lignification degree [41]. Food wastes are characterized by high levels of salt, grease, carbohydrates, and moisture [42, 43]. Because of the different elemental compositions and properties of raw materials, HAs have different characteristics [44].

Concentration of solids in fermentation medium at the disposable loading of the bioreactor has been increased and data correlate with the increase of amine nitrogen content that verifies biosynthetic processes (**Table 1**). According to dynamics of enzymatic activities, proteolytic activity of microorganism population decreases by 30 days and the increase of amine nitrogen could not be connected directly with the processes of proteolysis and obviously is the reflection of biosynthetic activity of microbial population. Data on the number of reducing sugars (their number reduction by 30–70 days) are logically kept within the assumption about the change of carbon source (transfer from utilization of easily hydrolyzed substrates to the using of hardly hydrolyzed) and, obviously at this stage of fermentation, the processes of cellulose hydrolysis have been intensified, that is, confirmed by the increase of cellulolytic activity of microbial population. On the basis of data obtained (humic

Sample	pH	D_{450}	Amine N (mg/ml)	Reducing agents (%)	HLS (%)	E_4/E_6*
PAF20	7.9	0.38	0.63	0.35	1.4	4.1
PAF30	7.6	0.52	0.89	0.33	1.5	7.1
PAF70	7.9	0.68	0.56	0.18	7.6	6.9
PAF100	8.1	1.35	1.12	0.34	8.4	6.8
PAF150	7.1	1.55	1.82	0.39	10.2	3.8
PAF200	6.8	3.50	2.73	0.50	13.1	11.0

**The E_4/E_6 ratio is considered to be inversely related to the degree of aromaticity of the humic substances and to their degree of humification*

Table 1.
Characteristics of PAF samples on different fermentation stages.

substance yield), it can be concluded preliminary that from 70-th day of fermentation, the stage of humic substances' synthesis starts.

Content of HLS formed in the process of fermentation has been increasing to the end of fermentation and partially correlated with the increase of solids, amine nitrogen, and reducing substances. However, the degree of humification significantly ranges on different stages, probably showing biosynthetic activity of microbial consortium on the accomplishment of the processes of "maturing" and "transformation" of humic compounds. The inclusion of the decomposition products of easily degradable compounds, i.e. sources of carbon and the products of cellulose hydrolysis determines the degree of humification corresponding to soil humic acids which contain significant amount of aliphatic fragments, carbohydrates, peptides, and small proteins. It is logical to assume that the reduction of this parameter from 7.1 to 6.8 reflects larger content of humic acids in the preparations as compared to fulvic acids, i.e., testifies about the increase of aromatic structures and the degree of condense. Sharp decrease of this parameter to 3.8 indicates about the production of highly condensed compound with high content of aromatics—actually about the formation of humic compound nucleus. From the other side, the presence of enzymatic activities, proteolytic, hydrolase, and cellulase in microbial population on these terms of fermentation, allows to assume the disintegration of aliphatic fragments, carbohydrates, and peptides that significantly reduce the determining value E_4/E_6. Increase of E_4/E_6 on the last stages of fermentation indicates the appearance of new synthesized macromolecules.

Preparations of HLS extracted from PAFs had a good solubility in low alkaline solutions at pH 12 and precipitated well from the solution at their acidification to pH 2; i.e., HLS demonstrate the properties of natural humic acids (HA) close to the class of microbial and soil humic substances in terms of element composition and spectral characteristics (**Table 2**; **Figures 1** and 2). Absorption spectra of alkaline solutions of the produced preparations in UV and visible field of spectra presented descending curves without specific strips of absorption (**Figure 1**). Significant absorption in UV field, reducing with the increase of wave length and small "shoulder" at wavelength 280 nm was specific for the studying preparations as well as for natural humic acids.

FTIR spectra allowed not only to assess qualitative composition of the functional groups but also to propose the model of their synthesis and transformation in the process of anaerobic fermentation. A wide intensive band of absorption at 3152–3270 cm^{-1} corresponds to valent oscillations of OH group (**Figure 2**). Small peaks in the field 2923–2924 cm^{-1} could be caused by the oscillations of the aliphatic groups CH_2 and CH_3. There is a small peak in the field 2587 cm^{-1} on the spectrum of PAF-150, obviously responsible for OH-group oscillations. Greatly expressed peak of absorption with maximum at 1667 cm^{-1} on PAF spectrum is a characteristic

Sample	Atom % in HLS						
	C	H	N	O	H/C	O/C	C/N
HLS PAF20	41.2	34.8	4.1	19.9	0.84	0.48	10.0
HLS PAF30	34.4	39.3	4.0	22.1	1.14	0.64	8.6
HLS PAF70	34.3	38.1	4.3	23.2	1.11	0.68	7.98
HLS PAF100	35.5	38.4	3.6	22.6	1.08	0.64	9.87
HLS PAF150	41.8	30.5	4.1	23.6	0.73	0.56	10.45
HLS PAF200	33.9	41.5	3.8	20.8	1.22	0.61	8.92

Table 2.
Elemental analysis of HLS.

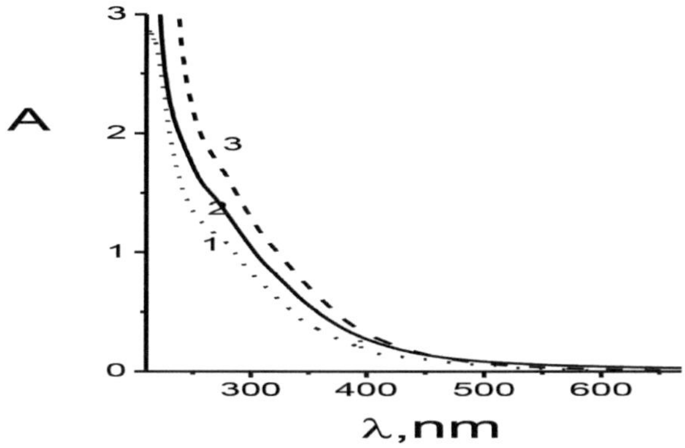

Figure 1.
Electron spectra of HLS: (1) soil HA; (2) HLS PAF150; (3) HLS PAF200.

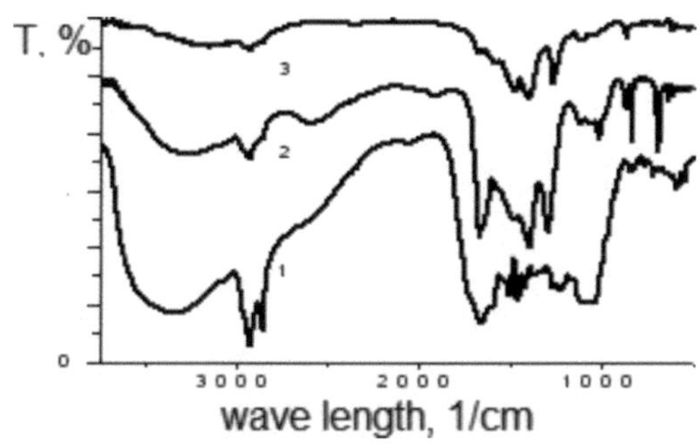

Figure 2.
FTIR spectra of HLS: (1) soil HA; (2) HLS PAF150; (3) HLS PAF200.

for double bonds C=C, —CH=CH_2, and =C=CH_2. For PAF 200 spectrum, the absorption in this field is expressed weaker. On both spectra, there is a peak at 1570–1577, responsible for the oscillation of C=N bond, and also of aromatic structures. The peak near 1470–1473 cm^{-1}, responsible for the oscillations of CH_2 and CH_3 groups, is distinctly presented on PAF 200 spectrum, to more extent than on PAF-150 spectrum. Distinct peak on both spectra in the field of 1396–1406 cm^{-1} can be attributed to the oscillations of COOH and —CHO groups. Absorption in this field is greatly expressed on both spectra. Primary and secondary alcohol groups could be responsible for the peak in the field 1270–1298 cm^{-1}. Absorption on both spectra in the field 1131 cm^{-1} can be linked with the oscillations of CO-alcohol and carbon groups. The availability of the above-mentioned atomic groups is an indication that the isolated preparations of humic substances are close to the other microbial humic-like substances and also to natural humic acids.

HLS of PAF presented two fractions according to size-exclusion chromatography analysis. One fraction goes out by sharp peak into the field of free volume of the column (it is also the characteristic for soil humic substances). It indicates that in all studied preparations there is high molecular fraction by molecular mass about 80 kD (determined from calibration curve). Low molecular fraction corresponds to about 5 kD by molecular wt. This peak was also sharp by form and differed from the wide peak on elution curve of soil humic substances with molecular weight about 23 kD.

Electrophoregrams of capillary electrophoresis typical for the HS from PAF samples (data are not shown) are similar to the electropherograms of the HA standards. Compounds migrating in 10–15 min interval and presented in the electropherograms as peaks with uneven front and extended end ("hump") are presented.

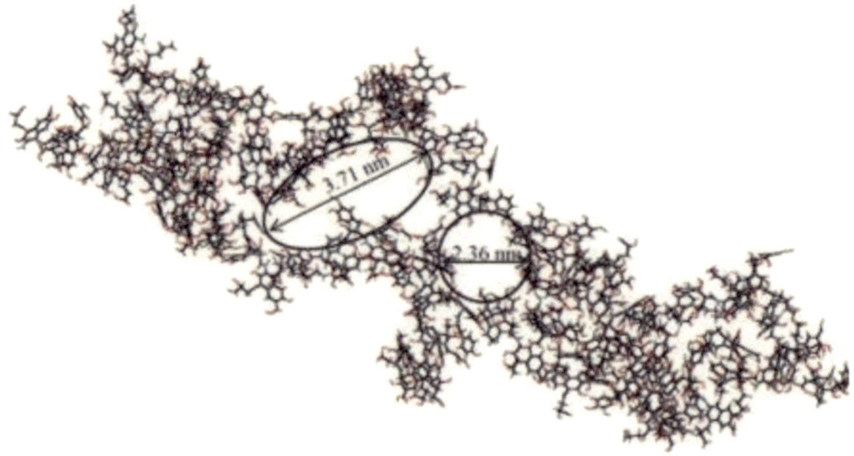

Figure 3.
Formula of humic substances [45].

The amount of the HS formed during a fermentation increases up to the end of a fermentation. The humification degree sharply changes at various stages.

According to Alvarez-Puebla et al.'s HS model [45], simple (though heterogeneous) monomeric units progressively build up into high-molecular weight polymers by random condensation and oxidation process (**Figure 3**). Accepting this model, we can assume the following mechanism of humic substance formation in the process of anaerobic fermentation: the first stage includes microbiological synthesis of humic substance nucleus, containing a great number of aliphatic fragments (initial period of fermentation). Next inclusion into synthesized nucleus of aromatic structures and/or the transformation of aliphatic fragments into aromatic structures takes place in the period between 30 and 150 days. The humic substances produced by their properties, spectral and spectroscopic characteristics, are mostly close to natural humic substances. The second stage includes further transformation of humic substances (150–200 days of fermentation) and can represent two processes: inclusion and/or formation of aromatic structures and hydrolysis of humic substances by microorganisms. As shown in [35], humic substances can play a role of electron acceptors for anaerobic respiration of microorganisms, as redox mediators for the processes of recovery and as donors of electrons for microorganisms. It should be noted that *Bacillus subtilis* can use quinone derivatives as donors of electrons and take part in the fermentation process and can participate in the transformation of humic substances. The third stage actually confirms the completion of high aromatic structure formed in the second stage and formation of aggregates. The stability of HS aggregates in solution is dynamic and influenced by solution ionic strength and pH.

2.4 Plant growth enhancement

Humic substances contribute to the growth and health of agricultural plants [46]. Moreover, HAs have been reported to have positive effects on the growth of wheats, ornamental plants, peas, and many other economically valued plants [47–49]. A recent random-effect meta-analysis of HS applied to plants [16] concluded on an overall dry weight increase of 22 ± 4% for shoots and of 21 ± 6% for roots. HSs can promote plant growth, which seems to be related to their positive influence on root architecture and the soil environment. Nardi et al. [50] revealed that HAs can promote the uptake of Na, Ba, and P in plants and modify the pH of the soil surrounding the root by stimulating the activity of H^+-ATPase in plant roots. Sharif et al. [51]

revealed that application of HAs to potted corn significantly increased root and shoot biomass, while Tahir et al. [47] found that the largest increase in plant height and shoot weight occurred when HAs were at a concentration of 60 mg/kg soil. In addition, HAs increase the cell membrane permeability, which can increase nutrient uptake and accumulation [47, 49]. In summary, the application of HSs can enhance the seed germination, rooting, seedling growth, and nutrient use of plants. HSs are therefore ideal for use in place of synthetic plant growth regulators.

According to a review by Jardin [15], humic substances have been recognized for long as essential contributors to soil fertility, acting on physical, physico-chemical, chemical, and biological properties of the soil. Most biostimulant effects of HS refer to the amelioration of root nutrition, via different mechanisms. One of them is the increased uptake of macro- and micronutrients, due to the increased cation exchange capacity of the soil containing the polyanionic HS, and due to the increased availability of phosphorus by HS interfering with calcium phosphate precipitation. Another important contribution of HS to root nutrition is the stimulation of plasma membrane H^+-ATPases, which convert the free energy released by ATP hydrolysis into a transmembrane electrochemical potential used for importing nitrate and other nutrients. Besides nutrient uptake, proton pumping by plasma membrane ATPases also contributes to cell wall loosening, cell enlargement, and organ growth [52]. HSs seem to enhance respiration and invertase activities providing C substrates. The proposed biostimulation activity of HS also refers to stress protection. Phenylpropanoid metabolism is central to the production of phenolic compounds, involved in secondary metabolism and in a wide range of stress responses. High-molecular mass HSs have been shown to enhance the activity of key enzymes of this metabolism in hydroponically-grown maize seedlings, suggesting stress response modulation by HS [53, 54].

It has been stated that digestates contain bioactive substances, such as phytohormones (e.g., gibberellins and indoleacetic acid), nucleic acids, monosaccharides, free amino acids, vitamins and fulvic acid, etc., with the potential to promote plant growth and to increase the tolerance to biotic and abiotic stress [55]. Digestates have higher contents of indoleacetic acid than the original plant feedstock [56]. This increase could only be explained by a microbial synthesis during the digestion process. The biotests of the identified fractions of HLS showed that all the fractions are active. But the efficiency of stimulating assay depends on the

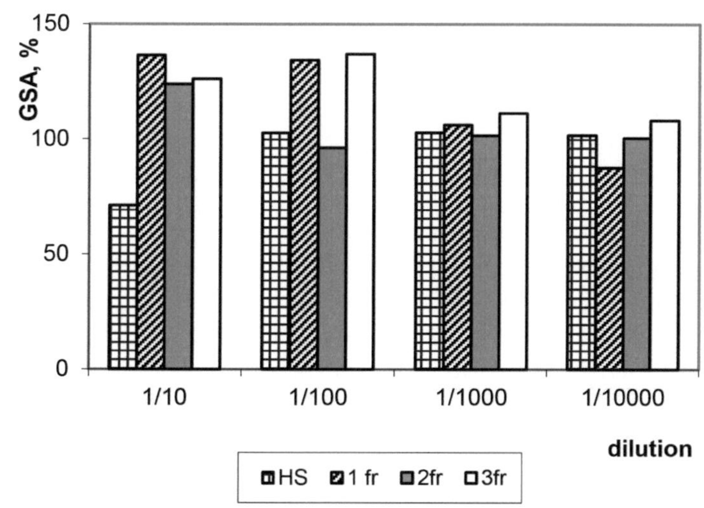

Figure 4.
Influence of HLS on the wheat seed germination: (1) PAF70; (2) PAF100; (3) PAF150 (GSA—growth-stimulating activity, 1fr—fraction with molecular weight ca. 5 kD, 2fr—ca. 50 kD, 3fr—ca. 100 kD).

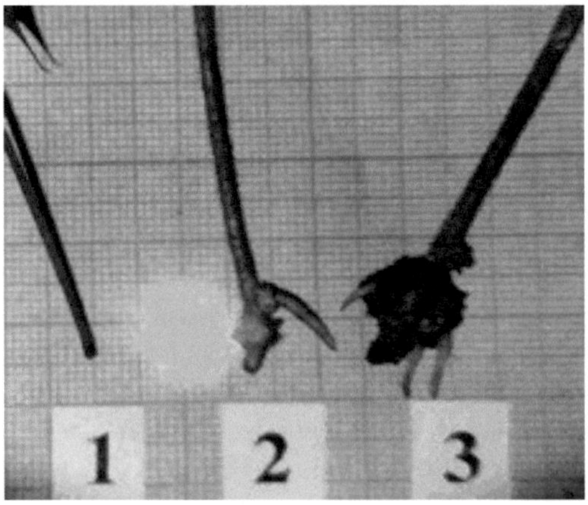

Figure 5.
Auxin biotest: (1) control; (2) 10^{-6} M IAA; (3) 0.1 g/L HLS PAF70 (the auxin-like activity of HLS was estimated using the tillers of liana sp. Cissus L. *of* Vitaceae. *The callus produced and root system of tillers was used as an auxin-test response. Tillers of liana were placed in water solution (1), in 10^{-6} M of IAA (indole-acetic acid) (2), and in 0.1 g/L of HLS of PAF70 (3). The rhizogenesis was observed for 60 days at 25°C).*

dilution degree. In our case, the specific biological activity of the separated fractions of HS has been determined (**Figure 4**). The effect of "mutual exclusion"of the specific biological activity has been established in case of the unfractionated HP.

The biological system test on the rhizogenesis of liana *Cissus* L. with a high IAA-oxidase activity toward the exogenetic auxins has shown an auxin-like effect of HA (**Figure 5**). The size of callus in the sample with HLS significantly increases such as in IAA sample. It can be assumed that there are auxins in the concentration 10^{-6}–10^{-7} M for HLS. Although hormonal effects are described, whether HSs contain functional groups recognized by the reception/signaling complexes of plant hormonal pathways, liberate entrapped hormonal compounds, or stimulate hormone-producing microorganisms is often unclear [57].

The stimulating effect of HP samples depended on metal ions, and in the field of low concentrations for any of the metal ions, the effect was higher than of model solution of metal ions (**Figure 6**). The increase of metal concentration leads to the increase of growth-stimulating activity of model solution. The increase of metal ions concentration has no clear stimulating effect of humic preparations. The HP inhibit the roots growth at the 10 times dilution.

The inhibiting effect of HLS of PAF to the germination of roots is identical to inhibition effect of the salted soils that may be caused by a high content of osmotic components (OC). The concentration of sodium ions in HLS of PAF-70-200 was 10–24 g/L, and in PAF-20 and PAF-30, it exceeded the limited value. Concentrations of K ions in all the samples also exceeded the detection limit. To differentiate the effect of metal ions and HA the GSA was compared for the following model solutions: (1) metal ions, (2) metal ions with HA, and (3) metal ions with HA and osmotic components (Na^+ and K^+). The effect of osmotic components on growth-stimulating activity is illustrated in **Figure 7**.

In all series of the experiment, a dose-dependent effect of growth-stimulating activity on the concentration of the components of model solutions and HLS of PAF-150 is observed. Comparative analysis of the curves shows that without OC the growth-stimulating activity of model solutions increased with the metal concentration and the presence of HS increased the the roots growth as compared to the metal solution. Addition of the OC to the medium for seed soaking caused the reduction of growth stimulation, even to the inhibition of seed germination.

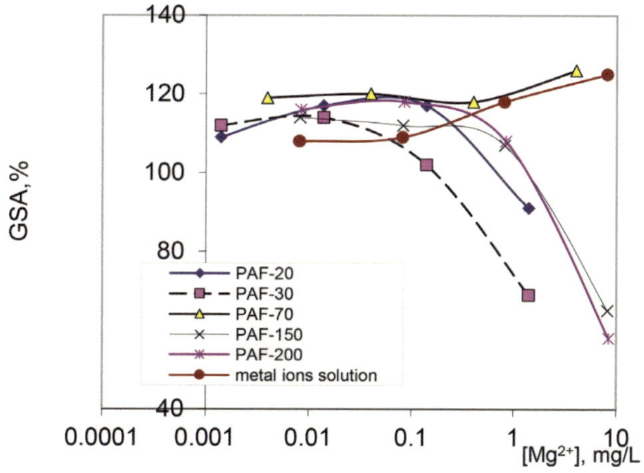

Figure 6.
*Effect of Mg^{2+} on HLS growth-stimulating activity (*Comparative biotesting. *Seedling technique was used. Seeds of wheat* Lada *were incubated for 3 days at 24 ± 1°C at constant illumination. As a control, HLS PAF-150 was used. Concentration of HLS was selected so as the optic density value of model solution corresponded to that in HLS PAF-150. pH of model solution of metal ions is 5.5. The length of roots was used as a test response).*

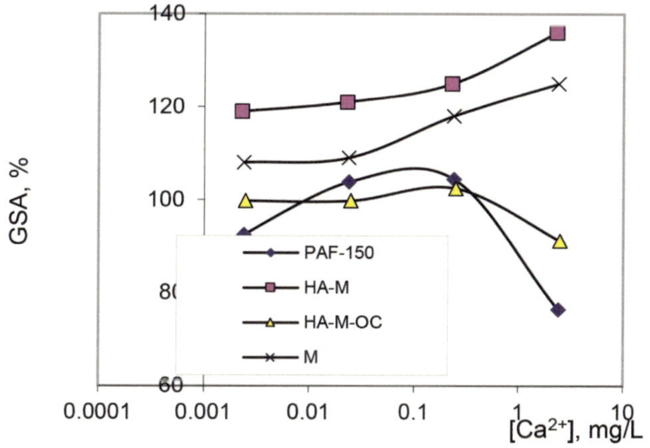

Figure 7.
Effect of Ca^{2+} on growth-stimulating activity of PAF-150 and HA (PAF150—products of anaerobic fermentation in 150 days, HA-M—humic acids with Ca^{2+}, HA-M-OC—humic acids with Ca^{2+} and osmotic components, M—Ca^{2+}).

2.5 Membrane filtration of the active fractions of humic preparations

The comparison of micro- and ultrafiltration processes allows to produce the preparations with different level of physiological activity (growth-stimulating activity, GSA). The process of microfiltration allows to produce in the permeate the final fraction that stimulates the growth of the roots approximately for 60%. The observing dynamics of an increase in growth-stimulating activity in the permeates allows to make a conclusion about the degradation of humic complex at the reduction of ionic strength of the solution and/or the removal of metals from the complexes of metal-humic substances. As a result of such degradation, we observe the appearance of the fractions with higher physiological activity, and this tendency is a characteristic not only for the permeates but also for the concentrates. The developed technological approach gives a possibility to separate the solid phase from the target product of the fermentation not only for one stage but also to increase significantly its physiological activity. Besides, this approach can be recommended for the production of low molecular fractions of humic substances of different origin, which are rather prospective from one side, for the study of their effect to membrane transport in plant and microbial cells, and from the other side, it could be used as active additives to the applied organic-mineral fertilizers.

3. Conclusion

The basic biological active components of liquid anaerobic fermentation by-products are the substances of the humic-like nature. Although humic substances may be useless for methane production during anaerobic digestion, they are useful raw materials for organic fertilizers. Common commercial humic fertilizer is primarily produced from peat, brown coal, and weathered coal. When compared with commercial humic substances, sludge humic substances contain a wider variety of organic substances, more lipids, more nitrogen, and a lesser degree of oxidation [2]. The process of anaerobic fermentation of the mixed wastes of plant and animal origin reflects the dynamics of microbial population change and humic-like by-product evolution. The content and composition of humic-like substances on different stages of fermentation are varied. By behavior in alkali, acids, element composition, spectroscopic characteristics, data of capillary electrophoresis, and size-exclusion chromatography, the humic-like substances as by-products of organic waste anaerobiosis are close to the class of microbial and soil humic substances.

The level of growth-stimulating activity of humic-like substances depended on metal content in the products of anaerobic fermentation. The process of microfiltration to extract bioactive fractions could be used for the isolation of two types of fractions: one of them is the fractions close to natural humic substances, and the second is plant hormone-like substances as auxins. The ultrafiltration allows to remove the excess amount of osmotic components and, therefore, to increase growth-stimulating activity of the preparations. Concentrates as a depot of HLS produced using micro- and ultrafiltration can meet the requirements of long-term liquid fertilizers. Permeates can be used in different dilutions as extraroot feeding because of mineral components and active fractions of humic-like substances.

Conflict of interest

The authors declare no conflict of interest.

Author details

Sharipa Jorobekova[1] and Kamila Kydralieva[2*]

1 Institute of Chemistry and Phytotechnologies, Bishkek, Kyrgyzstan

2 Moscow Aviation Institute (National Research Institute), Lomonosov Moscow State University, Moscow, Russia

*Address all correspondence to: kamila.kydralieva@gmail.com

References

[1] Demuynck M, Nyns E, editors. Biogas Plants in Europe. A Practical Handbook. Netherlands: Springer; 1984. 339 p

[2] Li H, Li Y, Zou S, Li S. Extracting humic acids from digested sludge by alkaline treatment and ultrafiltration. Journal of Material Cycles and Waste Management. 2014;**16**(1):93-100. DOI: 10.1007/s10163-013-0153-6

[3] Moller K, Muller T. Effects of anaerobic digestion on digestate nutrient availability and crop growth: A review. Engineering in Life Sciences. 2012;**12**(3):242-257. DOI: 10.1002/elsc.201100085

[4] Da Ros C, Cavinato C, Pavan P. Optimization of thermophilic anaerobic digestion of winery bio-waste by micro-nutrients augmentation. Environmental Engineering and Management Journal. 2015;**14**:1535-1542. DOI: 10.30638/eemj.2015.165

[5] Batstone DJ, Keller J, Angelidaki I, Kalyuzhnyi SV, Pavlostathis SG, Rozzi A, et al. The IWA anaerobic digestion model no. 1 (ADM1). The 9th World Congress on anaerobic digestion. 02-06 Sep 2002;**45**:65-73. DOI: 10.2166/wst.2002.0292

[6] Gerardi MH. The Microbiology of Anaerobic Digesters. 1st ed. Somerset, New Jersey: Wiley Press; 2003. 177 p

[7] Appels L, Baeyens J, Degrève J, Dewil R. Principles and potential of the anaerobic digestion of waste-activated sludge. Progress in Energy and Combustion Science. 2008;**34**:755-781. DOI: 10.1016/j.pecs.2008.06.002

[8] Hernandez T, Moreno JI, Costa F. Characterization of sewage sludge humic substances. Biological Wastes. 1998;**26**:167-174. DOI: 10.1016/0269-7483(88)90163-2

[9] Wilén BM, Paul Lant BJ. The influence of key chemical constituents in activated sludge on surface and flocculating properties. Water Research. 2003;**37**:2127-2139. DOI: 10.1016/S0043-1354(02)00629-2

[10] Dignac MF, Urbain V, Rybacki D, Bruchet A, Snidaro D, Scribe P. Chemical description of extracellular polymers: Implication on activated sludge floc structure. Water Science and Technology. 1998;**38**:45-53. DOI: 10.1016/S0273-1223(98)00676-3

[11] Li H, Jin Y, Nie Y. Extraction and utilization of humic acid from sewage sludge. Journal of Tsinghua University. 2009;**49**:1980-1983

[12] Stevenson FJ. Humus Chemistry: Genesis, Composition, Reactions. 2nd ed. New York: John Wiley and Sons; 1994

[13] Kulikowska D. Kinetics of organic matter removal and humification progress during sewage sludge composting. Waste Management. 2016;**49**:196-203. DOI: 10.1016/j.wasman.2016.01.005

[14] Ayuso M, Moreno JL, Herna'ndez T, Garcı'a C. Characterisation and evaluation of humic acids extracted from urban waste as liquid fertilisers. Journal of the Science of Food and Agriculture. 1997;**75**:481-488. DOI: 10.1002/(SICI)1097-0010(199712)75

[15] du Jardin P. Plant biostimulants: Definition, concept, main categories and regulation. Scientia Horticulturae. 2015;**196**:3-14. DOI: 10.1016/j.scienta.2015.09.021

[16] Rose MT, Patti AF, Little KR, Brown AL, Jackson WR, Cavagnaro TR. A meta-analysis and review of plant-growth response to humic substances: Practical implications for agriculture.

In: Sparks DS, editor. Advances in Agronomy. Vol. 124. USA: Elsevier; 2014. pp. 37-89. DOI: 10.1016/B978-0-12-800138-7.00002-4

[17] Hwang KY, Shin EB, Choi HB. A mechanical pretreatment of waste activated sludge for improvement of anaerobic digestion system. Water Science and Technology. 1997;**36**:111-116. DOI: 10.1016/S0273-1223(97)00731-2

[18] Chiu YC, Chang CN, Lin JG, Huang SJ. Alkaline and ultrasonic pretreatment of sludge before anaerobic digestion. Water Science and Technology. 1997;**36**:155-162. DOI: 10.1016/S0273-1223(97)00681-1

[19] Vlyssides AG, Karlis PK. Thermal-alkaline solubilization of waste activated sludge as a pre-treatment stage for anaerobic digestion. Bioresource Technology. 2004;**91**:201-206. DOI: 10.1016/S0960-8524(03)00176-7

[20] Takashima M, Tanaka Y. Application of acidic thermal treatment for one- and two-stage anaerobic digestion of sewage sludge. Water Science and Technology. 2010;**62**: 2647-2654. DOI: 10.2166/wst.2010.490

[21] Chi Y, Li Y, Fei X, Wang S, Yuan H. Enhancement of thermophilic anaerobic digestion of thickened waste activated sludge by combined microwave and alkaline pretreatment. Journal of Environmental Sciences. 2011;**23**:1257-1265. DOI: 10.1016/S1001-0742(10)60561-X

[22] Vigueras-Carmona SE, Ramı'rez F, Noyola A, Monroy O. Effect of thermal alkaline pretreatment on the anaerobic digestion of wasted activated sludge. Water Science and Technology. 2011;**64**:953-959. DOI: 10.2166/wst.2011.726

[23] Saha M, Eskicioglu C, Marin J. Microwave, ultrasonic and chemo-mechanical pretreatments for enhancing methane potential of pulp mill wastewater treatment sludge. Bioresource Technology. 2011;**102**:7815-7826. DOI: 10.1016/j.biortech.2011.06.053

[24] Lin Y, Wang D, Wu S, Wang C. Alkali pretreatment enhances biogas production in the anaerobic digestion of pulp and paper sludge. Journal of Hazardous Materials. 2009;**170**:366-373. DOI: 10.1016/j.jhazmat.2009.04.086

[25] Navia R, Soto M, Vidal G, Bornhardt C, Diez MC. Alkaline pretreatment of Kraft mill sludge to improve its anaerobic digestion. Bulletin of Environmental Contamination and Toxicology. 2002;**69**:869-876. DOI: 10.1007/s00128-002-0140-4

[26] Li H, Jin Y, Nie Y. Application of alkaline treatment for sludge decrement and humic acid recovery. Bioresource Technology. 2009;**100**:6278-6283. DOI: 10.1016/j.biortech.2009.07.022

[27] Feng HJ, Hu LF, Mahmood Q, Long Y, Shen DS. Study on biosorption of humic acid by activated sludge. Biochemical Engineering Journal. 2008;**39**:478-485. DOI: 10.1016/j.bej.2007.11.004

[28] Li H, Li Y, Li C. Evolution of humic substances during anaerobic sludge digestion. Environmental Engineering and Management Journal. 2017;**16**(7):1577-1582. DOI: 10.30638/eemj.2017.171

[29] Zhang X, Zhou S, Zhou L. Component changes of humic substances and heavy metal distribution before and after sewage sludge composting. Chinese Journal of Ecology. 2004;**23**:30-33

[30] Bartoszek M, Polak J, Sukowski WW. NMR study of the humification process during sewage sludge treatment. Chemosphere. 2008;**73**:1465-1470. DOI: 10.1016/j.chemosphere.2008.07.051

[31] Amir S, Jouraiphy A, Meddich A, El Gharous M, Winterton P, Hafidi M. Structural study of humic acids during composting of activated sludge green waste: Elemental analysis, FTIR and ^{13}C NMR. Journal of Hazardous Materials. 2010;**177**:524-529. DOI: 10.1016/j.jhazmat.2009.12.064

[32] Polak J, Sułkowski WW, Bartoszek M, Papiez W. Spectroscopic studies of the progress of humification processes in humic acid extracted from sewage sludge. Journal of Molecular Structure. 2005;**744-747**:983-989. DOI: 10.1016/j.molstruc.2004.12.054

[33] Filip Z, Pecher W, Bertheli J. Microbial utilization and transformation of humic acid-like substances extracted from a mixture of municipal refuse and sewage sludge disposed of in a landfill. Environmental Pollution. 2000;**109**:83-89. DOI: 10.1016/S0269-7491(99)00229-8

[34] Zhu Y, Zhao Y. Stabilization process within a sewage sludge landfill determined through both particle size distribution and content of humic substances as well as by FT-IR analysis. Waste Management and Research. 2010;**29**:379-385. DOI: 10.1177/0734242X10384309

[35] Lovley DR, Coates JD, Blunt-Harris EL, Phillips EJP, Woodward JC. Humic substances as electron acceptors for microbial respiration. Nature. 1996;**382**:445-448. DOI: 10.1038/382445a0

[36] Lovley DR. Potential for anaerobic bioremediation of BTEX in petroleum-contaminated aquifers. Journal of Industrial Microbiology & Biotechnology. 1997;**18**:75-81. DOI: 10.1038/sj.jim.29002

[37] Bradley P, Chapelle FH, Lovley DR. Humic acids as electron acceptors for anaerobic microbial oxidation of vinyl chloride and dichloroethene. Applied and Environmental Microbiology. 1998;**64**:3102-3105

[38] Cervantes FJ, Dijksma D, Duong-Dac T, Ivanova A, Lettinga G, Field JA. Anaerobic mineralization of toluene by enriched sediments with quinones and humus as terminal electron acceptors. Applied and Environmental Microbiology. 2001;**67**:4471-4478. DOI: 10.1128/AEM.67.10.4471-4478.2001

[39] Cai L, Gao D, Chen TB, Liu HT, Zheng GD, Yang QW. Moisture variation associated with water input and evaporation during sewage sludge bio-drying. Bioresource Technology. 2012;**117**:13-19. DOI: 10.1016/j.biortech.2012.03.092

[40] Wu S, Shen Z, Yang C, Zhou Y, Li X, Zeng G, et al. Effects of C/N ratio and bulking agent on speciation of Zn and Cu and enzymatic activity during pig manure composting. International Biodeterioration and Biodegradation. 2016;**119**:429-436. DOI: 10.1016/j.ibiod.2016.09.016

[41] Wang X, Cui H, Shi J, Zhao X, Zhao Y, Wei Z. Relationship between bacterial diversity and environmental parameters during composting of different raw materials. Bioresource Technology. 2015;**198**:395-402. DOI: 10.1016/j.biortech.2015.09.041

[42] Kumar M, Ou YL, Lin JG. Co-composting of green waste and food waste at low C/N ratio. Waste Management. 2010;**30**:602-609. DOI: 10.1016/j.wasman.2009.11.023

[43] Zhou Y, Selvam A, Wong JW. Evaluation of humic substances during co-composting of food waste, sawdust and Chinese medicinal herbal residues. Bioresource Technology. 2014;**168**:229-234. DOI: 10.1016/j.biortech.2014.05.070

[44] Diallo MS, Simpson A, Gassman P, Faulon JL, Johnson JJH, Goddard WA III,

et al. 3-D structural modeling of humic acids through experimental characterization, computer assisted structure elucidation and atomistic simulations. 1. Chelsea soil humic acid. Environmental Science & Technology. 2003;**37**:1783-1793. DOI: 10.1021/es0259638

[45] Alvarez-Puebla RA, Goulet PJG, Garrido JJ. Characterization of the porous structure of different humic. Colloids and Surfaces A: Physicochemical and Engineering Aspects. 2005;**256**:129-135. DOI: 10.1016/j.colsurfa.2004.12.062

[46] Traversa A, Loffredo E, Gattullo CE, Senesi N. Water-extractable organic matter of different composts: A comparative study of properties and allelochemical effects on horticultural plants. Geoderma. 2010;**156**:287-292. DOI: 10.1016/j.geoderma.2010.02.028

[47] Tahir MM, Khurshid M, Khan MZ, Abbasi MK, Kazmi MH. Lignite-derived humic acid effect on growth of wheat plants in different soils. Pedosphere. 2011;**21**:124-131. DOI: 10.1016/S1002-0160(10)60087-2

[48] Zhang L, Sun X, Tian Y, Gong X. Biochar and humic acid amendments improve the quality of composted green waste as a growth medium for the ornamental plant *Calathea insignis*. Scientia Horticulturae. 2014;**176**:70-78. DOI: 10.1016/j.scienta.2014.06.021

[49] Maji D, Misra P, Singh S, Kalra A. Humic acid rich vermicompost promotes plant growth by improving microbial community structure of soil as well as root nodulation and mycorrhizal colonization in the roots of *Pisum sativum*. Applied Soil Ecology. 2017;**110**:97-108. DOI: 10.1016/j.apsoil.2016.10.008

[50] Nardi S, Pizzeghello D, Gessa C, Ferrarese L, Trainotti L, Casadoro G. A low molecular weight humic fraction on nitrate uptake and protein synthesis in maize seedlings. Soil Biology and Biochemistry. 2000;**32**:415-419. DOI: 10.1016/S0038-0717(99)00168-6

[51] Sharif M, Khattak RA, Sarir MS. Effect of different levels of lignitic coal derived humic acid on growth of maize plants. Communications in Soil Science and Plant Analysis. 2002;**33**:3567-3580. DOI: 10.1081/CSS-120015906

[52] Jindo K, Sonoki T, Matsumoto K, Canellas L, Roig A, Sanchez-Monedero MA. Influence of biochar addition on the humic substances of composting manures. Waste Management. 2016;**49**:545-552. DOI: 10.1016/j.wasman.2016.01.007

[53] Olivares FL, Aguiar NO, Rosa RCC, Canellas LP. Substrate biofortification in combination with foliar sprays of plant growth promoting bacteria and humic substances boosts production of organic tomatoes. Scientia Horticulturae. 2015;**183**:100-108. DOI: 10.1016/j.scienta.2014.11.012

[54] Schiavon M, Pizzeghello D, Muscolo A, Vaccaro S, Francioso O, Nardi S. High molecular size humic substances enhance phenylpropanoid metabolism in maize (*Zea mays* L.). Journal of Chemical Ecology. 2010;**36**:662-669. DOI: 10.1007/s10886-010-9790-6

[55] Liu W, Yang Q, Du L. Soilless cultivation for high-quality vegetables with biogas manure in China: Feasibility and benefit analysis. Renewable Agriculture and Food Systems. 2009;**24**:300-307. DOI: 10.1017/S1742170509990081

[56] Yu F, Lou X, Song C, Zhang M, Shan S. Concentrated biogas slurry enhanced soil fertility and tomato quality. Acta Agriculturae Scandinavica, Section B: Plant Soil Science. 2010;**60**:262-268. DOI: 10.1080/09064710902893385

[57] du Jardin P. The Science of Plant Biostimulants—A Bibliographic Analysis. Ad hoc Study Report to the European Commission DG ENTR. 2012. Available from: http://ec.europa.eu/enterprise/sectors/chemicals/files/fertilizers/final report bio 2012 en.pdf [Accessed: 19 June 2019]

Chapter 5

Compost Tea Quality and Fertility

Gaius Eudoxie and Micah Martin

Abstract

The water extract of compost termed "compost tea" retains all the beneficial soluble bioactive components, making it a potent source of plant stimulatory and defensive compounds. The exact nature and extent of these features are, however, modified by composting system, feedstock quality, tea preparation and resultant use and management, including application dynamics of the compost tea. Compost teas contain a significant quantity of total nutrients with the majority being primary macronutrients. Secondary and micronutrient concentrations are more variable, but contents are generally insufficient to satisfy crop requirements. Noting this, compost tea use in agriculture and horticulture supports crop nutrition directly and indirectly. Improvements in soil quality have been widely reported for a range of soils and compost teas. A key feature of compost tea-amended soils is the increase in soil organic matter and microbial diversity and its associated benefits. Research on appropriates rates for field and container use show large variability associated with edapho-climatic factors and crop species. However, foliar application seems best suited to maximising the dual nutrition and phytopathogenic effects of compost tea. Regardless of the purpose of compost tea use, its positive effects on crop growth and soil fertility, whilst controlling pest and disease, make it a contemporary sustainable tool aligned to organic agriculture.

Keywords: compost tea, substrate quality, soil fertility, crop nutrition, phytostimulation

1. Introduction

Composting and vermicomposting as bioengineering processes provide dual environmental and human health benefits through the reuse and recycling of biodegradable waste as well as the production for use of a biochemically stable compost and vermicompost, respectively. Thermophilic compost and vermicompost compared to other organic amendments sequester mineral nutrients whilst allowing release in a slow to controlled manner. The material once matured is pathogen free and phytotoxicity free, containing phytostimulatory biocompounds including humic substances, plant growth regulators (PGR) and other biomolecules that have been proven to enhance crop growth and development [1]. Doan et al. [2] reported better performance of vermicompost relative to compost and manure over a 3-year corn trial. They concluded that the combination of vermicompost and biochar further improved corn yield and water availability. The superior effects of vermicompost were also reported by Goswami et al. [3] who reported a substantial improvement in soil health and nutrient availability, physical stability and microbial diversity due to compost and vermicompost application. They noted interestingly that heavy metal contamination was less significant in vermicompost-treated soils.

Research evidence supports the positive influence of composts and vermicomposts as amendments and substrate components on plant performance and soil fertility and quality. Martínez-Blanco et al. [4], Doan et al. [2], Hubbe et al. [5] and Erhart and Hartl [6] reviewed the effects of composts on plant and soil properties and summarised key mechanisms by which the positive effects may be manifested. Erhart and Hartl [6] noted that the most important benefit of using compost is the increased soil organic matter content. This assertion is functionally more important under tropical humid climates where degradation is accelerated. Composts directly supply plant-available macronutrients, with lesser amounts of micronutrients depending on substrate and composting system and conditions. Indirectly either active biocompounds or soil microbial enhancement stimulates plant performance. Improvements and sustained plant health via suppression of disease agents also add to the list of secondary mechanisms.

Compost application to soil whilst effective in addressing soil fertility and soilborne diseases is limited in responding to direct plant diseases and acute nutritional deficiencies. A derivative of compost, compost tea, provides an opportunity to expand the reach and benefit of soluble compost components. Zaccardelli et al. [7] defined compost tea as an organic liquid product derived through extraction with water from quality compost carrying useful microorganisms and moieties capable of protecting and stimulating the growth of plants. Moieties include essential nutrients that can correct nutritional deficiencies during crop production. Compost teas have been shown to contain bioactive concentrations of extractable compounds of composts. The range and concentration of extractable compounds are dependent on compost substrates, composting method and extraction protocols [8]. Xu et al. [9] studied the effects of increasing aeration on compost tea chemical properties and reported increased nutrient contents and humification with increasing aeration. Islam et al. [8] reported significant decreases in total N, organic C and organic matter and slight decreases of bacterial and fungal communities of compost teas with increasing extraction time to 6 days. Contrastingly, Hegazy et al. [10] reported increasing microbial populations with increasing extraction time although exceeding 48 h did not show significant improvement. They also found decreased microbial concentration with dilution and temperatures above 28°C. In another study on compost tea properties, Remedios Morales-Corts et al. [11] stated greater nutritional content, humic acids, salicylic acid and indole acetic acid (IAA) for aerated compost tea vs. aerated vermicompost tea, but the effect on tomato performance was non-significant.

Compost tea has evolved to include variants such as leachates, washes and extractants from other organic sources (e.g. manure) and biodynamic preparations. The application and use of compost teas in organic systems require technical clarification to ensure that the quality associated with the amendment cum fertiliser is not compromised. One key distinction is the controlled ratio of compost to water which separates compost teas from organic leachates and washes. Additionally, the incorporation of additives including molasses to boost microbial activity may or may not improve compost tea properties and provide positive effects on plant and soil health. Palmer et al. [12] inoculated compost teas with *Escherichia coli* (*E. coli*) at the start of the extraction and noted that when teas were supplemented with 1% molasses, there was a significant increase in *E. coli*. The same was not true for treatments devoid of molasses.

Non-supplement compost teas have been shown to serve as a nutrient source, improving soil fertility and crop nutrition through direct and indirect mechanisms. Taha et al. [13] observed that application of compost tea significantly increased soil bacteria (including N_2 fixing) and fungi populations along with increasing N, P and K uptake of radish leaves by 150, 90 and 253%, respectively, compared to the control. These findings were supported by Mohd Din et al. [14] who reported non-consequential effects between aerated and non-aerated extraction of compost tea on pak choi performance. Increasing yields and plant performance traits from foliar and soil application of compost teas were further reported for pepper [7], navel orange [15], tomato [16] and cucumber [17]. The occurrence of combined suppressive and biostimulatory mechanisms, sustained by microbial communities, nutrient supply and carbon-based bioactive compounds, is assumed to underlie the positive effects [16]. Molecular characterisation through NMR suggests that supramolecular organic structures contained in compost tea may be associated with the biostimulatory response [16].

The duality of compost teas makes them ideally suited as organic fertilisers with the advantage of stimulating further organic nutrient release from inherent soil organic matter. This chapter covers the nature and behaviour of this nutrient source and provides an analysis of effects and mechanisms. It intends to showcase the range of possibilities of compost teas, even to the domain of hydroponics [17], which has been dominated by inorganic nutrient sources.

2. Definitions and standards

The term "compost tea" when used in the literature covers a wide range of aqueous solutions and/or suspensions made from different organic materials via a range of processes. The lack of consensus in definition and standards supports the need for clarification and distinction in reporting on compost tea, especially where claims are purported. St. Martin [18] reported that the terms compost tea and compost extract are typically used interchangeably but argued that they should be differentiated. Scheuerell and Mahaffee [19] defined compost extracts as the filtered products of composts mixed with any solvent (usually water), but not fermented or brewed, whilst Litterick et al. [20] defined compost tea as the filtered product of compost fermented in water. As such, tea differs from extract based on steeping period, with tea associated with a significantly longer brewing time. For either extract or tea, specific concentrations are prepared primarily on a mass to volume basis. This prescribed dilution allows distinction between compost leachate and tea. Zhou et al. [21] stated that composting leachate is a complex type of organic waste water originating during the composting process as a result of the constant application of water to maintain the substrate moisture content in the range of 65–70% [22]. The fact that compost leachate is collected throughout the process further distinguishes it from compost tea, which is an extraction product of mature compost. It is expected that compost leachate may present greater phytotoxic concern. The concentration of compost leachate also remains a mystery limiting standardisation.

Compost leachate, extract and tea can be produced from either vermicompost or thermophilic compost. Edwards et al. [23] noted that the superior biochemical and physical properties of vermicompost over thermophilic compost are reflected in tea quality and subsequent plant growth, although other factors including concentration, substrate, additives and brewing methods modify the response [24]. Reported differences in the performance of vermicompost and thermophilic compost may warrant specific terminology. Vermicompost usually causes little to no phytotoxic effect on plants; therefore, it is ready for use following harvesting. Remedios Morales-Corts et al. [11] reported vermicompost tea having an EC value that was approximately fivefold lower than thermophilic compost tea from the same feedstock, inferring lower potential phytotoxicity. Thermophilic compost tends to require a prolonged mesophilic (curing) phase to ensure proper humification and reduction in phytotoxicity. The production of compost tea for use as an organic fertiliser would require greater attention to final compost quality. Hegde et al. [25] specifically mentioned that vermiwash (worm tea) contained a mixture of excretory products and mucus secretions of earthworms along with essential nutrients. Such differentiation is also important as other organic amendments apart from composts are used to produce "teas". Zarei et al. [26] compared the quality of vermicompost tea versus vermiwash from different compost sources including leaf meal and cow manure. Results indicated a significantly greater nutrient content for a 1:10 m/v aerated vermicompost tea brewed for 24 h.

Manure tea is another variant of the concept of compost tea, made from a solution that contains animal manure. Azeez et al. [27] defined manure tea as the liquid extract from manure or a solution made by soaking manure in water in order to ease the decomposition process and enhance the release of nutrients. Although similar processes may be used, there is concern over the possible presence of pathogens in manure tea [28]. There are also commercially available microbial sources such as "effective microorganisms" that may replace compost as an inoculant, if teas are primarily aimed at increasing microbial content. The practice of fortifying compost and their tea extracts has grown among composters. Some of the common additives used include kelp, molasses, fish hydrolysate, rock dust, humic acid and carrot juice along with biodynamic extracts of plants such as "comfrey" [24, 29, 30]. The primary aim of this practice is to stimulate beneficial microbes by enriching their environment with a source of food and oxygen with the expectation of increased crop protection and a general boost in plant health. Naidu et al. [30] reported a 10–100-fold increase for total bacteria, fungi and actinomycetes for enriched compost tea relative to the control (pure compost tea). The authors also further reported that the enriched compost tea remained stable for up to 4 months. On the contrary, the use of kelp and molasses favoured the growth of *E. coli* in spiked compost teas. Strict sanitation practices can therefore act as a preventative measure to compost tea contamination by pathogens. It is noteworthy that several authors, Palmer et al. [12], Kannangara et al. [31] and Brinton et al. [29], reported the absence or minor presence of human pathogens in matured composts and their teas. Carrot juice acted as an inhibitor to the proliferation of *E. coli* in compost tea [31]. The use of additives during the brewing process has a lesser effect on tea mineral nutrient content. Compost maturity is a stronger determinant of tea nutritional quality [32]. A mature compost refers to decomposed organic matter that has no phytotoxic effects on plants [33]. Griffin and Hutchinson [34] stated that mature compost generally releases higher levels of soluble nutrients and fewer phytotoxic organic acids and heavy metals than immature material. The nutrient composition of finished compost is based on initial substrate quality as well as compost maturity. Immature composts can negatively affect compost tea quality and encourage anaerobic conditions [24, 29]. Careful management of the factors controlling the composting process is therefore important in ensuring an effective compost tea is derived. Observing some key maturity indices such as C/N ratio, pH and EC can help improve decision-making on final compost quality.

Similar to the distinctions previously discussed, which focused on organic material source, compost tea may be extracted under aerated or non-aerated conditions. During aerated extraction, air is pumped through water containing compost to maintain the O_2 level above 5 mg l^{-1} [24]. St. Martin [18] noted that most of the reported literature does not identify the O_2 concentration during preparation of aerated compost tea (ACT), which may imply that wide variability exists in process standards. For passive extraction, compost is placed in a certain volume of water and allowed to steep for several days, with or without occasional stirring [35]. The term fermentation has been strongly associated with the production of non-aerated compost tea (NCT) as a consequence of the presumed anaerobic conditions. However, St. Martin [18] explained that the term "brewing" is better suited to describe the process and implies a steeping process of compost in any solvent, which lasts for more than 1 h [36]. Considering the product and the process, a more appropriate terminology might be "compost infusion". Both terms, brewing and tea, imply the use of hot water [37]. Kevin [36] provides the best definition of compost tea which encapsulates the previous discussion. Compost tea is a commercially and anecdotally popularised term for an "infusion" where compost is steeped in water for a period of time with the aim of transferring soluble organic matter, beneficial microbes and nutrients in solution.

Greater effort is required by the scientific community and compost enthusiasts to define the product and process when referring to compost infusions and related extracts. This would improve standardisation and clarity among researchers and users.

3. Compost tea as a nutrient source

3.1 Growth, development, crop nutritional status and soil fertility

Relative to their use for disease suppression or control, research into the use of compost teas and other bio-fertilisers (phytohormones and humic substances) as sources of plant essential nutrients is limited. This fact is more evident when considering effects on field-grown crops [30]. **Table 1** identifies a short but comprehensive list of research reports focused on the effects of compost tea on plant performance characteristics, crop nutritional status and soil fertility. Special attention was directed to distinguishing among treatment effects that will assist in better understanding the reported variability. Most research used either non-aerated or aerated compost teas, whilst other studies reported on manure or microbial teas, whilst still a few others used commercial mixtures labelled as compost teas. Both thermophilic compost and vermicompost were used and compared (in a few studies). Application methods, rates and frequencies as well as crops, season and soils varied across studies, making any general inference anecdotal at most.

Seminal works by Hargreaves et al. [38] investigating the comparative use of compost teas, soil-incorporated compost of the same substrates and inorganic fertiliser showed similar responses in soil fertility and plant nutritional status across treatments. Compost tea applied at 150–300 ml/2.4 m^2 weekly as a foliar spray resulted in leaf tissue N and K content of strawberries similar to plants treated with soil-incorporated compost and mineral fertiliser. N and K tissue content further correlated well in both trial years to mineralised N. At rates applied NCT was able to maintain plant nutrient status within the sufficiency range. However, the authors were concerned over the low soil test K levels under compost tea. In a previous study, Hargreaves et al. [39] reported similar nutrient-supplying potential of foliar-applied NCT on raspberry growth and development and nutrition. Lower tissue K

Compost feedstock	Brewing method/ concentration*	Crop plant	Summarised effects	Reference
Not specified	ACT, conc. not specified	Canada yew	Increases in plant biomass similar for compost tea and inorganic fertiliser. Fertiliser resulted in greater biomass allocation to shoots	Smith et al. [40]
Agro-waste compost	ACT and NCT, 10% m/v	Pak choi	Increased yield, mineral nutrient content and antioxidant levels in leaves for NCT + mineral fertiliser	Mohd Din et al. [14]
Cow dung, urine and palm sugar, milk	NCT and ACT Drench application not specified. Foliar 10%	Broccoli	Inclusion of compost tea, especially at higher doses, increased crop yield and quality. Compost + compost tea maintained soil fertility similar to mineral fertiliser	Sanwal et al. [41]
Cow manure, molasses (MT) Commercial EM	ACT and NCT 5% m/v and v/v for MT and EM, respectively	Collard and spinach	Application of MT or EM did not improve short-term yield or soil fertility	Knewtson et al. [28]
Ruminant compost MSWC	NCT 10% m/v	Strawberries	Compost teas provided equivalent levels of nutrients to strawberries compared with inorganic fertiliser Soil K levels decreased with compost tea application	Hargreaves et al. [42]
Ruminant compost MSWC	NCT 10% m/v	Strawberries	Compost teas did not improve the fruit quality and total antioxidant capacity compared to inorganic fertiliser	Hargreaves et al. [43]
MSWC	ACT 20% m/v	Strawberries	Foliar application of compost teas provided sufficient nutrients to strawberries for growth and resulted in equal yield compared to compost applications to soil	Hargreaves et al. [38]
Chicken manure vermicompost	ACT 5 and 10% m/v	Pak choi	Greatest plant response was observed with 5 and 10% tea Magnitude of response was greater under chicken manure fertilisation	Pant et al. [32]

Compost feedstock	Brewing method/ concentration*	Crop plant	Summarised effects	Reference
Not specified	Not specified	Borage	Higher rate of compost tea significantly increases plant growth and productivity	Ezz El-Din and Hendawy [44]
Manure compost	ACT 10% m/v	Water spinach	Under aquaponic conditions, yield of water spinach was significantly increased via foliar application	Bethe et al. [45]
MSWC	NCT 20% m/v	Brussels sprouts	Yields of all treatments were similar to where conventional fertiliser was used. Tea may only be useful as a supplemental fertility source applied to foliage	Radin and Warman [37]
Ruminant compost MSWC	NCT 10% m/v	Raspberries	Yield, total antioxidant capacity of fruit and vitamin C content were not affected by treatment. Compost tea supplied less K to raspberries compared to compost	Hargreaves et al. [39]
Not specified	Brew not specified 25% m/v	Basil	Compost tea in conjunction with compost improved vegetative growth and essential oils but decreased N content	Khalid et al. [46]
Not specified	Not specified	Sugar beet	Sugar yield and juice quality of increased with compost tea. Combination of compost tea with mineral N fertiliser further increased yield and quality properties	El-Gizawy et al. [47]
Farm refuse compost	ACT 10% w/v	Pomegranate	Foliar application gave greater leaf mineral and pigment content, yield and fruit parameters than soil drenching	Fayed [48]
Cow manure vermicompost	Direct leachate used	Sorghum	Leachate can be used directly without dilution. Stimulated plant development but NPK was required for maximum growth	Gutiérrez-Miceli et al. [49]
Not specified	Direct leachate 50% v/v	Curry leaf	Foliar spray increased yield over the control, especially when combined with fertiliser and manure application	Hedge et al. [25]

Compost feedstock	Brewing method/ concentration*	Crop plant	Summarised effects	Reference
Rice straw and animal manure	Brew not specified 3% m/v	Onion	Foliar application increased yield and quality when used in combination with mineral N	Mahmoud et al. [50]
Palm fruit and oil effluent (ME)	ACT, 20%	Muskmelon	Fertigation + foliar compost tea significantly increased fresh weight, TSS and other quality indices	Naidu et al. [30]
Tomato, escarole residue and wood chips	ACT, 20%	Tomato	Largest yields were showed by compost tea-managed plants, significantly exceeding those of control plots	Pane et al. [16]
Not specified	Not specified	Peppers, cucumber and sweet corn	Treatment did not affect responses	Russo and Fish [51]
Agricultural waste compost	Not specified	*Atriplex* and mesquite	Compost tea increased morphological traits and further improves soil fertility when combined with compost	Shourije et al. [52]
Cow dung, vegetable waste and 1:2 ratio mixture	Direct leachate	Strawberries	Foliar application increased marketable fruit yield of firmer fruit with improved quality attributes. Also reduced physiological disorders like albinism and malformation	Singh et al. [53]
Town refuse compost	ACT, 5% m/v	Radish	Compost tea + mineral N increased growth and yield and improved soil biochemical properties and microbial population	Taha et al. [13]
Not specified	Not specified	Cucumber	Under hydroponic production compost tea had significantly lower yield but higher antioxidant content	Santiago-López et al. [17]

*****ACT, aerated compost tea; NCT, non-aerated compost tea; MT, manure tea; EM, effective microbes; ACTME, aerated compost tea with microbial enhancer; MSWC, municipal solid waste compost; ME, microbial-enriched.*

Table 1.
Summary of studies examining the effects of compost tea on crop performance and nutrient status.

content in 2 out of 3 growth years with associated lower soil K test levels suggested a lower uptake potential of K through plant leaves as K concentration in the compost teas was fairly high. The inability of foliar-applied compost teas to supplement soil available nutrient levels may pose productivity risks, especially where foliar uptake may be hindered by morphological or biochemical mechanisms. Consideration should be given to increasing frequency of application. Hargreaves et al. [42] also reported increased Na content of raspberry leaves attributed to enhanced foliar uptake. This may be an issue for compost tea foliar use in Na-sensitive plants. Lazcano and Domínguez [1] noted that the use of vermicompost tea reduces the probability of phytotoxic effects arising from elevated EC and specific ion concentrations. However, variations in compost tea quality will determine the nature and extent of plant effects.

Radin and Warman [37] reported that foliar application of compost tea showed similar pH and soil K levels to other treatments, except for higher K under organic fertiliser. Doubling the application rate of compost tea had no significant effect on tissue nutrient content and soil fertility. The lower application rate of 450 ml/5.3 m^2, similar to [38, 39], resulted in comparable P, K, Ca and S contents. Tissue Na content was similar across compost tea treatments. However, these were significantly lower than the incorporated compost treatment, which contrast findings from [42]. Both studies utilised municipal solid waste compost (MSWC) from the same source. This highlights the extent of variability reported in compost tea effects and provides a rationale to continued interest in this area. In a study on Canada yew, Smith et al. [40] reported comparable biomass as well as morphological traits (plant height) between the biodynamic compost tea and inorganic fertiliser. The authors concluded that compost tea was capable of producing similar growth characteristics with lower nutrient input than inorganic fertiliser. They noted that compost tea influenced partitioning and allocation of biomass with a favourable allocation to roots relative to shoots. The opposite was observed for inorganic fertiliser. Naidu et al. [30] suggested that biodynamic compost tea can result in improved microbial populations and stability. Mahmoud et al. [50] and Pant et al. [54] reported enhanced overall root development accompanied with better nutrient uptake by tea-treated plants than mineral fertiliser. Although Smith et al. [40] did not report tissue nutrient contents, others have indicated similar tissue nutrient contents between compost tea and mineral fertiliser [37, 39], suggestive of either favourable root allocation of nutrients or positive effects of other tea bioactive components affecting biomass accumulation. Lazcano and Dominguez [1] suggested that improvements in plant growth and productivity seem less related to the amount of nutrients available, as plant tissue nutrient content varies non-significantly between doses. Other bioactive components of compost teas clearly stimulate and enhance plant performance beyond nutrient availability. Suggestions of the presence and action of phytohormones and humic substances have been reported [55].

Compost tea use as the sole or primary nutrient source may provide adequate nutrients to maintain plant growth and development, but that depend on application rate and frequency, strength (concentration) and crop species. Santiago-López et al. [17], in one of the only studies utilising compost tea as a hydroponic fertigation solution, reported, significantly lower yields for cucumber fertigated with either compost or vermicompost tea relative to Steiner's solution. The lower yields were reflective of lower nutrient contents of the teas and, probably more critically, imbalances among nutrient concentrations. The literature supports foliar application, which seems to have a positive effect on nutrient uptake. Relative to its use as a primary nutrient source, application of compost tea in conjunction with other fertilisers or plant additives results in a superior performance. Pant et al. [32] investigated the effects of increasing dilution of compost tea on pak choi grown on soil amended with organic or inorganic fertiliser. The addition of vermicompost tea increased pak choi height as well as base diameter across all fertiliser sources. N accumulation in plant leaves followed a similar trend. This study differed in respect to nutrient uptake compared to trials where compost tea was the primary nutrient source. Similar increases in plant nutrient content, especially N, have been reported [14] with analogous explanations put forward. The influence of compost tea on N availability and uptake is an important effect that has not been fully understood and evaluated. Reviewed studies failed to investigate N use efficiency under compost tea application, an area that has tremendous global significance. Keeling et al. [56] noted with interest that compost water extracts had no effect on shoot and root development of field bean compared to oilseed rape. This finding suggests that constituents of compost tea influence N uptake and possibly assimilation and may respond to chemical equilibrium. Further, Khan et al. [57] investigating the combinatory effects of compost tea and AMF stated that nutrient stoichiometry revealed that there was a greater uptake of N under vermicompost tea than P, whilst the opposite occurred with AMF.

It is notable that compost tea application resulted in positive effects when combined with either soil-incorporated mineral fertiliser or compost across crop species. For borage [44], basil [46], sugar beet [47], curry [25], onion [50] and peppers, cucumbers and sweet corn [51], supplementation with foliar-applied compost or vermicompost tea improved aboveground morphological traits, plant nutrient content and soil fertility above control treatments. This effect was in most cases more apparent when compost tea was applied in soils amended with compost. The literature on compost tea as a nutrient source speculates that compost tea improves nutrient use efficiency. However, there has not been any study that directly measured nutrient uptake and use efficiency. Different methods have traditionally been employed in investigating nutrient effects, but these are limited by priming effects [58]. Stable isotope techniques or other direct measurement protocols need to be employed to verify the effects of compost tea on nutrient uptake and use efficiency.

Many microbial water-based extracts are commercially available or can be made following simple instructions. These extracts are sometimes not differentiated in the literature from compost teas. Such terms, which may lead to misconception, are discussed in the preceding section on definitions and standards. Reports of neutral or negative effects of compost tea on plant growth and nutrition all had one common thread. Water extracts (not brewed) were made from organic sources other than compost. Knewtson et al. [28] investigated the effects of manure tea and commercial EM on collard greens and spinach as well as soil biological properties. Results of this study showed that tea application had non-significant effects across fertiliser source for plant biomass and soil biochemical properties, applied at rates and frequencies corresponding to previous reports showing positive benefits. Sanwal et al. [41] using a homemade microbial tea (made mainly from manure) reported no improvement in the use of compost tea over compost-only treatments, with both treatments outperforming synthetic fertiliser. Broccoli height, number of leaves, weight of leaves, head weight and diameter were all similar across compost and compost + compost tea treatments. No information was reported about the application methodology or the tea properties limiting an objective assessment. Variability in the depth of reporting of compost tea properties and application methodologies limits any meta-analysis or objective evaluation of results on plant growth and development. Greater attention needs to be paid by authors and publishers to ensure that such relevant information is provided.

3.2 Yield, yield components and quality

Single application of compost teas failed to result in a similar positive effect of increasing crop yield and quality likened to the effects on growth parameters [37–39]. Radin and Warman [37] reported lower but significantly similar yields of Brussels sprouts for foliar-applied compost tea than organic fertiliser and

MSWC. Hargreaves et al. [39], Hargreaves et al. [42] and Hargreaves et al. [43] using a similar MSWC all reported similar yields for strawberries and raspberries across compost tea and compost treatments. Both Radin and Warman [37] and Hargreaves et al. [42] further highlighted that yields were below average with plants showing visible nutrient deficiency symptoms. The former authors inferred through correlation analysis that reduced tissue nutrient concentrations may have resulted in lower yields for tea treatments and concluded that compost tea as a nutrient source does not provide sufficient plant nutrition. Remarkably, Hargreaves et al. [42] used rates 4–8 times greater than Radin and Warman [37] and at twice the frequency, with analogous results, although in this case compost tea values were similar to compost treatments. Contrastingly, Singh et al. [53] working on strawberries showed an increase in yield, fruit nutrient content and quality attributes compared to the control. However, the control was a no-nutrient water control. A notable difference between Singh et al. [54] and Radin and Warman [37] and Hargreaves et al. [42] is the use of vermicompost tea versus thermophilic compost tea. Whilst composts can be defined from a quality perspective in terms of stability and maturity, the processes resulting in the production of vermicompost and compost are vastly different, inferring differences in their bioactive components [59] and overall effects on crop performance.

Khalid et al. [46] and Ezz El-Din et al. [44] both investigated the combinatory effects of increasing compost tea concentration together with compost or fertiliser, respectively, on herb yields and quality. Inclusion of compost tea resulted in a significantly greater yield than the control, with the extent of increase directly related to increasing rate of compost tea supplementation [44]. Both studies showed that the concentration of essential oils and flavonoids [46] increased significantly for compost tea treatments. Increased yield has been correlated to improved growth characteristics and plant nutrient contents. However, the question remains; what mechanisms are responsible for these responses, especially where inorganic sources are applied? Are fertiliser nutrients temporarily immobilised or transformed, reducing potential loss? Or are they quickly absorbed by enhanced root systems? What is certain is that compost tea cannot supply the requisite amount of essential nutrients under field production systems; hence, the bulk of nutrients originate from the primary source when combined.

Works of El-Gizawy et al. [47] on sugar beet support the previous findings. Increased sugar yields and juice quality were aligned with increasing frequency of foliar-applied compost tea. The relative increase in sugar yield ranged from 6.5% for a one-time soil drench application to 36% with soil drenching followed by three monthly applications of compost tea. Juice quality measured by purity, sugar content and K content was significantly higher in plants treated with medium N (75 kg N/fad) and the highest frequency of compost tea. Sifola and Barbieri [60] reported that organic N sources and the combination with inorganic sources significantly improved plant height, root and shoot dry weight and oil yield of basil compared to inorganic N only. Hegde et al. [25] and Mahmoud et al. [50] showed identical findings for curry leaves and onion yield and oil content respectively. The combination of mineral fertiliser + organic manure + compost tea resulted in significantly higher yield attributes.

Akin to earlier discussed neutral and reduced effects of compost tea on plant performance, these studies reported the same for crop yield and quality components. Russo and Fish [51] investigated a commercial compost tea (PMSLA and EO-12) on peppers, cucumbers and sweet corn, Knewtson et al. [28] studied effects of commercial EM and manure tea on collard greens and spinach, and Sanwal et al. [41] investigated a traditional fresh cow manure-based tea on broccoli. Similar yields were reported across crops for comparative treatments with and without compost tea. Only for Sanwal et al. [41], who employed a fertiliser only control, did compost tea treatments show significantly higher yield and quality, but the effect was lessened by statistically similar values to compost-only treatments. A critical evaluation deciphering underlying mechanisms by which compost tea improves plant performance and specifically nutrient use efficiency will considerably add to better synthesis and use of compost tea.

4. Mechanisms of action of compost tea as nutrient sources

Similar to compost, compost tea is a microbiologically active, nutrient-rich extract, which when used to irrigate crops (foliar or soil drench) influences growth, yield, nutrition and quality directly or indirectly through chemical and/or biological mechanisms. Direct modalities involve increased nutrient supply and action of microbial bioactive compounds including humic acids and phytohormones. Indirect mechanisms operate principally on the effect of microorganisms within the compost tea on pest suppression and enhancement of microbial communities that affect direct mechanisms of nutrient uptake or production of bioactive compounds. **Table 2** provides a synthesis of reported studies elucidating mechanisms of action of compost teas.

4.1 Direct mechanisms

4.1.1 Nutrient content

Analysis of compost tea has revealed varying concentrations of plant mineral elements based on compost source, brewing methods and dilution. Pant et al. [32] investigated the effects of compost tea strength on pak choi growth and yield and reported that increasing vermicompost tea concentrations linearly and positively influenced plant growth resulting from increased concentration of mineral nutrients. Increasing amounts of available nutrients in compost tea and their relationship with crop growth have also been confirmed by [42, 61]. It has been postulated that the increased presence of soluble mineral nutrients can enhance nutrient uptake from soil and increase foliar uptake of nutrients [32]. However, this effect seems conjunctive with other chemical and possibly biological components within compost tea. Mahmoud et al. [50] claimed that the availability of mineral nutrients is greater for foliar versus drench applications. They inferred that compost tea increased the time stomata stay open, reducing loss from the leaf surface. Schönherr [62] identified polar aqueous pores which facilitate the absorption of charged ions into epidermal cells. This capability was further confirmed and explained by Kaya et al. [63] who stated that compost tea increased permeability of cellular membranes in plants to minerals which increased plant growth. In addition to the direct effects that foliar feeding has on nutrient assimilation, Fayed [48] asserted that it also has a positive priming effect and may actually promote root absorption of the same and other nutrients. Of the few studies that investigated compost tea use across different soils, Pant et al. [54] showed that soil properties modify nutrient absorption under compost tea fertilisation, with poor aeration and drainage limiting nutrient uptake. Increased nutrient uptake through compost tea has been reported to increase leaf area, which relatedly improves light interception, photosynthesis, water and nutrient use and dry matter production [64]. The importance of compost tea nutrient content to growth was reported by [49]. The authors showed that vermicompost tea explained ~50% of the growth parameters for sorghum and significantly affected total macronutrient content

Compost feedstock	Brewing method/ concentration	Compost tea properties	Summarised mechanisms	Reference
Grape pomace and manure biodynamic compost	Brewed for 8 h (aeration not confirmed) 5% m/v)	Nutrients	Direct effect of nutrients but more plausible effect of phytohormones, even though not determined	Reeve et al. [61]
Garden waste compost and vermicompost	ACT/AVT 5% v/v	Nutrients, humic acids, phytohormones, heavy metals, pathogens	Nutrient composition, humic acids, the presence of phytohormones	Remedios Morales-Corts et al. [11]
Ruminant compost MSWC	NCT 10% m/v	Nutrients	Positive effect associated with nutrient content	Hargreaves et al. [42]
Chicken manure vermicompost	ACT, NCT and augmented ACT 10% m/v	Nutrients, microbial content	Microbial and hormonal contributions along with nutritional effects	Pant et al. [54]
MSWC	ACT 20% m/v	Nutrient	Less effective in building SOM, nutrient availability, but insufficient N	Hargreaves et al. [38]
Chicken manure compost and vermicompost Green waste compost Food waste vermicompost	ACT 5 and 10% m/v	Macronutrients, humic acids, phytohormones	Positive effects largely associated with mineral N and GA_4	Pant et al. [65]
Not specified	Not specified	Nutrients and microbial content	Supply of nutrients and microbial function	Ezz El-Din and Hendawy [44]
MSWC	NCT 20% m/v	Nutrients and heavy metals	Low N availability affected P uptake inferring inadequate nutrient in tea	Radin and Warman [37]
Ruminant compost MSWC	NCT 10% m/v	Raspberries	Increased uptake of Na leading to nutrient toxicity	Hargreaves et al. [42]
Vegetable waste compost	ACT 25% v/v	Nutrients, microbial content	Physiological and nutritional biostimulation	Zaccardelli et al. [7]
Not specified	Brew not specified 25% m/v	Nutrients	Induced systemic response allowing for better growth	Khalid et al. [46]
Not specified	Not specified	Nutrients and bacterial content	Synergistic effect of beneficial microorganisms and essential micronutrients and other bioactive compounds	El-Gizawy et al. [47]

Compost feedstock	Brewing method/ concentration	Compost tea properties	Summarised mechanisms	Reference
Farm refuse compost	ACT 10% w/v	Nutrients	Nutrient availability and plant root physiological response increasing nutrient uptake	Fayed [48]
Fruit bunch and chicken manure compost	Brew not specified 20% m/v	Nutrients, microbial content	Beneficial microbes, presence of essential micronutrients and bioactive compounds	Siddiqui et al. [66]
Cow manure vermicompost	Direct leachate used	Nutrients, humic acids, microbial content	Specific mention of micronutrient stimulate in tandem with humic acids and phytohormones	Gutiérrez-Miceli et al. [49]
Rice straw and animal manure	Brew not specified 3% m/v	Nutrients, microbial content	Supply of nutrients and microbial function Asserted plant physiological effect increasing leaf nutrient absorption	Mahmoud et al. [50]
Palm fruit and oil effluent (ME)	ACT 20%	Nutrients and heavy metals	Availability of macro- and micronutrients and improved fertility of soilless media, presence of chelating agents	Naidu et al. [30]
Tomato, escarole residue and wood chips	ACT 20%	13C NMR, nutrients, Microbial content	Plant disease suppressiveness, supply of chelated nutrients and the action of humic acids and phytohormones	Pane et al. [16]
Agricultural waste compost	Not specified	Not reported	Mechanism not described	Shourije et al. [52]
Cow dung, vegetable waste and 1:2 ratio mixture	Direct leachate	Nutrients	Increased nutrient and growth regulator including humic acids	Singh et al. [53]
Biodynamic preparation (several components including compost)	Aerated ratio not specified	Chemical properties	Retention of root tip border cells which serves as bacterial trap and physical protection against pathogens	Tollefson et al. [67]

Compost feedstock	Brewing method/ concentration	Compost tea properties	Summarised mechanisms	Reference
SMC	ACT, 10% + K_H	Not reported	Improved nutrient availability and soil quality evidenced by greater CEC, OM and nutrient content	Taha et al. [13]
Rice straw compost and vermicompost, Cyprus bark compost	ACT, 10%	Chemical properties, microbial diversity	Foliar application increases speed and efficiency of nutrient uptake relative to soil drenching	Kim et al. [68]
Agro-waste compost	ACT and NCT, 10% m/v	Nutrients, microbial content	Interactive stimulatory effect of compost tea on uptake of mineral nutrients	Mohd Din et al. [14]

ACT, aerated compost tea; NCT, non-aerated compost tea; MT, manure tea; EM, effective microbes; ACTME, aerated compost tea with microbial enhancer; MSWC, municipal solid waste compost; ME, microbial-enriched; K_H, potassium humate; SMC, substrate mushroom compost.

Table 2.
Summary of reported mechanisms of action of compost tea on crop performance and nutrient status.

of plants. Due to the relatively small amount of nutrients in compost tea present in organic form, the possibility of slow release mechanisms is low. However, the presence of chelated nutrients increases availability to plants [44]. Complimentary to the supply of readily available essential nutrients, Hargreaves et al. [39] noted that compost tea produced from municipal solid waste compost (MSWC) contained high concentrations of Na^+ which when applied to leaves correlated with increased tissue Na content of raspberries. Whilst the concentration of beneficial mineral nutrients has been shown to increase through compost tea use, attention is warranted towards the presence of other soluble ions that may be harmful or toxic to plants. Characterisation of composts and compost teas should be a prerequisite to use, which can influence dilution and application rate and frequency.

4.1.2 Bioactive metabolites

There is a paucity of information on the chemical and biological properties of compost extracts, but many authors have identified and to some extent confirmed the presence of water-extractable mineral elements and biologically active metabolites such as humic acids and plant growth regulators (PGR). These latter compounds may enhance initial root development, nutrient uptake and plant growth. Keeling et al. [56] analysed compost tea by liquid chromatography mass spectrometry and identified the presence of hundreds of low (<20 kDa) molecular weight organic compounds, which may be involved in plant responses. Humic acids have been identified as an important component of compost teas, especially vermicompost teas. de Sanfilippo et al. [69] suggested that earthworm activity accelerated humification of organic matter and their influence in increasing microbial populations enhances the presence of humic acids. Humic acid stimulatory effect on plants has been explained by direct action, which is hormonal in nature, together with an indirect action on the metabolism of soil microbes and the uptake of soil nutrients by plants [70, 71]. This effect is greater on roots, resulting in increased proliferation of root hairs and enhancement of root initiation [70]. Valdrighi et al. [72] and Cacco et al. [73] reported increased N uptake associated with humic acids. The latter increases the permeability of membranes of root cells and or switch on NO_3^- transport genes in roots. There remains an open debate on the exact mechanisms of increased nutrient uptake with Panuccio et al. [74] suggesting that humic substances only stimulated NH_4^+ uptake. Consensus exists on the fact that there is activity at the cellular level. Keeling et al. [56] in their study on compost tea effects on field bean seedling performance provided strong support for the N transport mechanism effect. They reported that neither root nor shoot development was stimulated by compost tea, consistent with the notion that compost components modify transport of inorganic N compounds within roots and that N-fixing legumes are insensitive to the process. Humic acids have been also associated with increased stress tolerance and produced similar endogenous levels of osmoprotectants as exogenous levels of PGR [75].

Spaccini et al. [76] reported that ACT contained low molecular weight bioactive compounds of microbial origin. In addition to humic acids, phytohormones and other metabolites have been identified in compost teas. Arancon et al. [77] reported a small quantity (198 ng/L) of GA4 in a chicken manure-based compost tea, which resulted in significantly greater root growth. Garcia Martinez et al. [78] also found that compost aqueous extracts contained compounds with molecular structure and biological activity analogous to auxins. The range of concentrations of phytohormones in compost teas varies much like its mineral nutrient content and is reported to also have a concentration-based effect. Studies on pak choi by Pant et al. [65] showed nonquantifiable amounts of phytohormones in a range of compost teas but improved growth and yield in treatments receiving compost tea. Notwithstanding the probable effect of other bioactive compounds, it can be reasoned that only trace levels of PGRs are required to initiate a plant response. In the same study, the authors detected traces of GA4 in a mature thermophilic compost associated with specific fungal families. The specific nature of bioactive metabolic compounds in compost tea still remains unresolved. However, there is strong evidence that suggests the major effect is expressed on plant roots both at the cellular and phenotypic levels, which serves as a system for increased nutrient uptake. Further research into biochemical pathways triggered by isolated compounds from compost teas will aid in elucidating these mechanisms.

4.2 Indirect effects

4.2.1 Disease suppression

St. Martin [18] provides an extensive review of the effects and mechanisms, whereby compost tea suppresses microbial diseases, insect pests and other plant pathogens. Suppression has been attributed to direct suppression of pathogens or to the induction of systemic resistance. Other authors have suggested that improved plant health provides an indirect effect on plant growth and nutrition. Healthy plants with thick cuticles are better able to resist attack from piercing and biting insects as well as microbial infections. Hargreaves et al. [42] noted the complimentary effects of compost tea suppression of diseases on overall plant growth and nutrient uptake. Khalid et al. [46] working with basil stated that foliar-applied biotic extracts are believed to initiate a systematic response known as “induced resistance”, which may act as a repellent or reduce the severity of pest and disease. The nature of this induction is uncertain, but Ezz El-Din and Hendawy [44] inferred that foliar application of compost tea provides useful microbes that colonise leaf surfaces, which probably competes with pathogens.

The nature, diversity and concentration of microorganisms present in compost tea may influence its ability to suppress pathogens or inoculate the receiving plant with beneficial microbes. The concentration and diversity of microorganisms in compost tea differ and in most instances are lower relative to the compost source. Arancon et al. [79] determined that vermicompost teas had about 1/3 of the microbial activity and diversity of the solid vermicompost (v/v). However, there was little diminution of the influence on tomato and cucumber seedling growth over the trial period. No specific family of microbes present in compost teas has been shown to have a critical role in nutrient uptake and growth of treated plants. Compost tea-related changes in soil and tissue microbial biodiversity have been reported to increase the range of biocontrol agents [80] and increase the production of defensive substances by the plant [7]. Compost tea may also impart a physical, morphological defence in plant roots through altered dispersal of border cells, which have a high affinity to trap bacteria. Tollefson et al. [69] investigating several plant species noted retention of root border cells under compost tea treatment compared to water even at high agitation.

4.2.2 Microbial inoculants

Soil application of compost tea resulted in greater presence of N fixers, actinomycetes and spore formers [81]. This supports the previous discussion on compost tea positive effect on N availability and plant uptake. Carpenter-Boggs [82] contested that compost tea is thought to act more as a microbial inoculant that stimulates soil and foliar microbial production effectiveness than as a direct nutrient source. The inoculation potential of compost tea has not been extensively researched, and it seems less likely that reported concentrations of microorganisms applied at typical dilutions and rates would serve as inoculants dominating the diversity of complex microbial systems such as soil. Reeve et al. [61] contented that at a rate of 5 g per preparation per 11 mg material, it is unlikely that the teas are effective microbial inoculants. They suggested that a more plausible mode of action may be through hormonal action.

Whether compost teas provide sufficient microorganisms to inoculate soil or other growing media may not be as critical as their stimulatory effects on indigenous microbes. Natarajan [83] suggested that microbes present in compost tea compliment activities of native microbes favouring decomposition of organic matter at a faster rate, resulting in better transformation of nutrients and their availability to crops. This position is supported by greater respiration rates and dehydrogenase activity for compost tea-treated soils arising from greater availability of active organic carbon or enrichment of nutrients for the microbes through addition of high organic carbon content compost [84]. It is notable that this response is similar for soil amended with either mineral fertilisers or compost [66]. Sanwal et al. [41] investigated broccoli performance under different nutrient management systems and found lower yield for inorganic fertiliser than compost tea plus fertiliser. They concluded that organics would increase the retention and slow release of nutrients at critical periods of crop growth and improve microbial properties.

Current knowledge suggests that compost teas work through a combination of chemical and biological mechanisms, which have not been fully unravelled. A ready supply of macro- and chelated micronutrients becomes more available to plants through hormonal action of humic acids and other phytohormones that act both on the roots and leaves. The variability that exists across compost teas' chemical and biological constituents compounded by edaphic and crop factors challenges precise determination of mechanistic effects.

5. Conclusion

There are several organic fertilisers and nutrient sources available but with few liquid options. Compost tea presents the best alternative liquid organic nutrient source for horticultural and agricultural use. Its origin in compost ensures that the product is sanitary and contains soluble constituents of the compost. By definition, being associated with mature compost also minimises the potential for phytotoxic compounds and effects on crop health and soil quality. The term compost tea must be differentiated from other extracts and from other organic sources as these may have potential negative or non-stimulatory effects. Regardless of the nature of the composting system, composting feedstock and brewing conditions, compost tea has been reported to enhance soil quality through increased microbial diversity and nutrient availability and increase crop growth and importantly yield. The latter is especially so when compost tea is combined with mineral or organic fertilisers. Several mechanisms have been posited for the altered effects associated with compost tea use including increased availability and uptake of nutrients especially when applied as a foliar treatment. Secondary mechanisms include increased soil organic matter and nutrients turnover through microbial activity. Stimulatory effects occur on plants through PGRs, humic and other biostimulatory compounds present in compost teas. Further benefit is derived through the suppression of plant pathogens which provides the best opportunity for maximum growth. As an amendment its versatility betters even its source material. Compost tea has shown potential for being an ideal beneficial product in any cropping system.

Author details

Gaius Eudoxie* and Micah Martin
Department of Food Production, UWI, St. Augustine, Trinidad and Tobago

*Address all correspondence to: gaius.eudoxie@sta.uwi.edu

References

[1] Lazcano C, Domínguez J. The Use of Vermicompost in Sustainable Agriculture: Impact on Plant Growth. New York, USA: Nova Science Publishers, Inc; 2011. pp. 1-23

[2] Doan TT, Rumpel C, Janeau J-L, Jouquet P, Henry-Des-Tureaux T. Impact of compost, vermicompost and biochar on soil fertility, maize yield and soil erosion in Northern Vietnam: A three year mesocosm experiment. Science of the Total Environment. 2015;**514**:147-154. Available from: https://www.researchgate.net/publication/271837945

[3] Goswami L, Nath A, Sutradhar S, Bhattacharya SS, Kalamdhad A, Vellingiri K, et al. Application of drum compost and vermicompost to improve soil health, growth, and yield parameters for tomato and cabbage plants. Journal of Environmental Management. 2017;**200**:243-252. DOI: 10.1016/j.jenvman.2017.05.073

[4] Martínez-Blanco J, Lazcano C, Christensen TH, Muñoz P, Rieradevall J, Møller J, et al. Compost benefits for agriculture evaluated by life cycle assessment. A review. Agronomy for Sustainable Development. 2013;**33**(4):721-732

[5] Hubbe MA, Nazhad M, Sánchez C. Composting of lignocellulosics. BioResources [Internet]. 2010;**5**(4):2808-2854. Available from: https://bioresources.cnr.ncsu.edu/BioRes_05/BioRes_05_4_2808_Hubbe_NS_Composting_Review_1298.pdf

[6] Erhart E, Hartl W. Genetic engineering, biofertilisation, soil quality and organic farming. In: Lichtfouse E, editor. Genetic Engineering, Biofertilisation, Soil Quality and Organic Farming [Internet]. Sustainable Agriculture Reviews. Vol. 4. Dordrecht: Springer Netherlands; 2010. pp. 311-345. Available from: http://link.springer.com/10.1007/978-90-481-8741-6

[7] Zaccardelli M, Pane C, Villecco D, Maria Palese A, Celano G. Compost tea spraying increases yield performance of pepper (*Capsicum annuum* L.) grown in greenhouse under organic farming system. Italian Journal of Agronomy. 2018;**13**(3):229-234

[8] Islam MK, Yaseen T, Traversa A, Ben Kheder M, Brunetti G, Cocozza C. Effects of the main extraction parameters on chemical and microbial characteristics of compost tea. Waste Management [Internet]. 2016;**52**:62-68. DOI: 10.1016/j.wasman.2016.03.042

[9] Xu D, Zhao S, Xiong Y, Peng C, Xu X, Si G, et al. Biological, physicochemical, and spectral properties of aerated compost extracts: Influence of aeration quantity. Communications in Soil Science and Plant Analysis [Internet]. 2015;**46**(18):2295-2310. DOI: 10.1080/00103624.2015.1081693

[10] Hegazy MI, Hussein E, Salama A, Salama A. Improving physico-chemical and microbiological quality of compost tea using different treatments during extraction. African Journal of Microbiology Research. 2015;**13**(50):763-770

[11] Remedios Morales-Corts M, Pérez-Sánchez R, Ángeles Gómez-Sánchez M. Marcelo Gonçalves de Oliveira C. Efficiency of garden waste compost teas on tomato growth and its. Scientia Agricola [Internet]. 2017;**75**(5):400-409. DOI: 10.1590/1678-992X-2016-0439

[12] Palmer AK, Evans KJ, Brown J, Ross T, Metcalf DA, Palmer AK. Potential for growth of *E. coli* in aerobic compost extract. Compost Science & Utilization. 2010;**18**(3):152-161

[13] Taha SS, Seoudi OA, Abdelaliem YF, Tolba MS, El Sayed SSF. Influence of bio-spent mushroom compost tea and potassium humate as a sustainable partial alternate source to mineral-N influence of bio-spent mushroom compost tea and potassium humate as a sustainable partial alternate source to mineral-N fertigation on. Egyptian Journal of Basic and Applied Sciences. 2018;**33**(1):103-122

[14] Mohd Din ARJ, Cheng KK, Sarmidi MR. Assessment of compost extract on yield and phytochemical contents of Pak Choi (*Brassica rapa* cv. *chinensis*) grown under different fertilizer strategies. Communications in Soil Science and Plant Analysis. 2017;**48**(3):274-284. DOI: 10.1080/00103624.2016.1269793

[15] Omar AEDK, Belal EB, El-Abd AENA. Effects of foliar application with compost tea and filtrate biogas slurry liquid on yield and fruit quality of Washington navel orange (*Citrus sinenesis* Osbeck) trees. Journal of the Air & Waste Management Association. 2012;**62**(7):767-772

[16] Pane C, Palese AM, Spaccini R, Piccolo A, Celano G, Zaccardelli M. Enhancing sustainability of a processing tomato cultivation system by using bioactive compost teas. Scientia Horticulturae [Internet]. 2016;**202**:117-124. DOI: 10.1016/j.scienta.2016.02.034

[17] Santiago-López G, Preciado-Rangel P, Sánchez-Chavez E, Esparza-Rivera JR, Fortis-Hernández M, Moreno-Reséndez A. Organic nutrient solutions in production and antioxidant capacity of cucumber fruits. Emirates Journal of Food and Agriculture. 2016;**28**(7):518-521

[18] St. Martin CCG. Potential of compost tea for suppressing plant diseases. CAB Reviews Perspectives in Agriculture Veterinary Science Nutrition and Natural Resources. 2014;**9**(032):1-38. Available from: http://www.cabi.org/cabreviews/review/20153038604

[19] Scheuerell S, Mahaffee W. Compost tea: Principles and prospects for plant disease control. Compost Science & Utilization. 2002;**10**:313-338. Available from: http://apps.webofknowledge.com.proxy.library.cornell.edu/full_record.do?product=UA&search_mode=GeneralSearch&qid=120&SID=2DeKI57IjfNd2a7JkIB&page=1&doc=1

[20] Litterick AM, Harrier L, Wallace P, Watson CA, Wood M. The role of uncomposted materials, composts, manures, and compost extracts in reducing pest and disease incidence and severity in sustainable temperate agricultural and horticultural crop production—A review. Critical Reviews in Plant Sciences. 2004;**23**:453-479

[21] Zhou C, Wang R, Zhang Y. Fertilizer efficiency and environmental risk of irrigating impatiens with composting leachate in decentralized solid waste management. Waste Management. 2010;**30**(6):1000-1005. DOI: 10.1016/j.wasman.2010.02.010

[22] Tejada M, Gonzalez JL, Hernandez MT, Garcia C. Agricultural use of leachates obtained from two different vermicomposting processes. Bioresource Technology. 2008;**99**(14):6228-6232

[23] Edwards CA, Arancon NQ, Greytak S. Effects of vermicompost teas on plant growth and disease. Biocycle. 2006;**47**(5):28

[24] Ingham ER. The Compost Tea Brewing Manual. 5th ed. Soil Food Web Incorporated: Oregon; 2005. p. 79

[25] Hegde NK, Siddappa R, Hanamashetti SI. Response of curry leaf (*Murraya koenigii* Spreng) "suvasini" for foliar spray of vermiwash and nutritional treatments. Acta Horticulturae. 2012;**933**:279-284

[26] Zarei M. Jahandideh Mahjen Abadi VA, Moridi A. Comparison of vermiwash and vermicompost tea properties produced from different organic beds under greenhouse conditions. International Journal of Recycling of Organic Waste in Agriculture. 2018;**7**(1):25-32. DOI: 10.1007/s40093-017-0186-2

[27] Azeez JO, Ibijola TO, Adetunji MT, Adebisi MA, Oyekanmi AA. Chemical characterization and stability of poultry manure tea and its influence on phosphorus sorption indices of tropical soils. Communications in Soil Science and Plant Analysis. 2014;**45**(20):2680-2696

[28] Knewtson SJB, Griffin JJ, Carey EE. Application of two microbial teas did not affect collard or spinach yield. HortScience. 2009;**44**(1):73-78

[29] Brinton W, Storms P, Evans E, Hill J. Compost teas: microbial hygiene and quality. Biodynamics. Jan 2004;**2**:36-45

[30] Naidu Y, Meon S, Kadir J, Siddiqui Y. Microbial starter for the enhancement of biological activity of compost tea. International Journal of Agriculture and Biology. 2010;**12**(1):51-56

[31] Kannangara T, Forge T, Dang B. Effects of aeration, molasses, kelp, compost type, and carrot juice on the growth of escherichia coli in compost teas. Compost Science & Utilization. 2006;**14**(1):40-47

[32] Pant AP, Radovich TJK, Hue NV, Paull RE. Biochemical properties of compost tea associated with compost quality and effects on pak choi growth. Scientia Horticulturae. 2012;**148**:138-146. DOI: 10.1016/j.scienta.2012.09.019

[33] Wichuk KM, McCartney D. Compost stability and maturity evaluation—A literature review. Journal of Environmental Engineering and Science. 2013;**8**(5):601-620. Available from: http://www.icevirtuallibrary.com/doi/10.1680/jees.2013.0063

[34] Griffin TS, Hutchinson M. Compost maturity effects on nitrogen and carbon mineralization and plant growth. Compost Science & Utilization. 2007;**15**(4):228-236

[35] Weltzien HC. Biocontrol of foliar fungal diseases with compost extracts. In: Microbial Ecology of Leaves. New York, NY: Springer; 1991. pp. 430-450

[36] O'Rell K. NOSB Recommendation for Guidance: Use of Compost, Vermicompost, Processed Manure and Compost Tea. National Organic Standards Board Crops Committee Recommendation for Guidance Use of Compost, Vermicompost, Processed Manure, and Compost teas [Internet]. 2006. Available from: https://www.ams.usda.gov/sites/default/files/media/NOP Final Rec Guidance use of Compost.pdf

[37] Radin AM, Warman PR. Assessment of productivity and plant nutrition of Brussels sprouts using municipal solid waste compost and compost tea as fertility amendments. International Journal of Vegetable Science. 2010;**16**(4):374-391

[38] Hargreaves JC, Adl MS, Warman PR. Are compost teas an effective nutrient amendment in the cultivation of strawberries? Soil and plant tissue effects. Journal of the Science of Food and Agriculture. 2009;**89**(3):390-397

[39] Hargreaves J, Adl MS, Warman PR, Rupasinghe HPV. The effects of organic amendments on mineral element uptake and fruit quality of raspberries. Plant and Soil. 2008;**308**(1-2):213-226

[40] Smith RF, Cameron SI, Letourneau J, Livingstone T, Livingstone K, Sanderson K. Assessing the Effects of Mulch, Compost Tea, and Chemical Fertilizer on Soil Microorganisms, Early Growth, Biomass Partitioning,

and Taxane Levels in Field-grown Rooted Cuttings of Canada Yew (*Taxus canadensis*). [Internet]. 2006. Available from: https://cfs.nrcan.gc.ca/publications?id=32892

[41] Sanwal SK, Laxminarayana K, Yadav DS, Rai N, Yadav RK. Growth, yield, and dietary antioxidants of broccoli as affected by fertilizer type. Journal of Vegetation Science. 2006;**12**(2):13-26

[42] Hargreaves JC, Adl MS, Warman PR, Rupasinghe HPV. The effects of organic and conventional nutrient amendments on strawberry cultivation: Fruit yield and quality. Journal of the Science of Food and Agriculture. 2008;**88**(15):2669-2675

[43] Hargreaves JC, Adl MS, Warman PR, Warman PR. The effects of municipal solid waste compost and compost tea on mineral element uptake and fruit quality of strawberries. Compost Science & Utilization. 2009;**17**(2):85-94

[44] Ezz El-Din AA, Hendawy SF. Effect of dry yeast and compost tea on growth and oil content of borago officinalis plant. Research Journal of Agriculture and Biological Sciences. 2010;**6**(4):424-430

[45] Bethe LA, Salam MA, Fatema UK, Rana KS. Effects of molasses and compost tea as foliar spray on water spinach (*Ipomoea aquatica*) in aquaponics system. International Journal of Fisheries and Aquatic Studies. 2017;**5**(3):203-207

[46] Khalid KA, Hendawy SF, El-Gezawy E. *Ocimum basilicum* L. production under organic farming. Research Journal of Agricultural and Biological Sciences. 2006;**2**(1):25-32

[47] El-Gizawy E, Shalaby G, Mahmoud E. Effects of tea plant compost and mineral nitrogen levels on yield and quality of sugar beet crop. Communications in Soil Science and Plant Analysis. 2014;**45**(9):1181-1194

[48] Fayed TA. Effect of compost tea and some antioxidant applications on leaf chemical constituents, yield and fruit quality of pomegranate. World Journal of Agricultural Sciences. 2010;**6**:402-411. Available from: http://www.idosi.org/wjas/wjas6(4)/9.pdf

[49] Gutiérrez-Miceli FA, García-Gómez RC, Rincón Rosales R, Abud-Archila M, María Angela OL, Cruz MJG, et al. Formulation of a liquid fertilizer for sorghum (*Sorghum bicolor* (L.) Moench) using vermicompost leachate. Bioresource Technology. 2008;**99**(14):6174-6180

[50] Mahmoud E, El-Gizawy E, Geries L. Effect of compost extract, N2-fixing bacteria and nitrogen levels applications on soil properties and onion crop. Archives of Agronomy and Soil Science. 2014;**61**(2):185-201

[51] Russo VM, Fish WW. Efficacy of microbial amendments on vegetables in greenhouse and field trials. HortScience. 2012;**47**(3):349-355

[52] Ansar Shourije F, Sadeghi H, Pessarakli M. Effects of different types of composts on soil characteristics and morphological traits of two dry rangeland species. Journal of Plantnutrition. 2014;**37**(12):1965-1980

[53] Singh R, Gupta RK, Patil RT, Sharma RR, Asrey R, Kumar A, et al. Sequential foliar application of vermicompost leachates improves marketable fruit yield and quality of strawberry (Fragaria × ananassa Duch.). Scientia Horticulturae. 2010;**124**(1):34-39

[54] Pant A, Radovich TJ, Hue NV, Arancon NQ. Effects of vermicompost tea (aqueous extract) on pak choi

yield, quality, and on soil biological properties. Compost Science & Utilization. 2011;**19**(4):279-292

[55] Atiyeh RM, Lee S, Edwards CA, Arancon NQ, Metzger JD. The influence of humic acids derived from earthworm-processed organic wastes on plant growth. Bioresource Technology. 2002;**84**(1):7-14. Available from: http://www.ncbi.nlm.nih.gov/pubmed/12137272

[56] Keeling AA, McCallum KR, Beckwith CP. Mature green waste compost enhances growth and nitrogen uptake in wheat (*Triticum aestivum* L.) and oilseed rape (*Brassica napus* L.) through the action of water-extractable factors. Bioresource Technology. 2003;**90**(2):127-132

[57] Khan MH, Meghvansi MK, Gupta R, Veer V, Singh L, Kalita MC. Foliar spray with vermiwash modifies the arbuscular mycorrhizal dependency and nutrient stoichiometry of bhut jolokia (*Capsicum assamicum*). PLoS One. 2014;**9**(3):e92318

[58] Gouveia GA, Eudoxie GD. Distribution of fertiliser N among fixed ammonium fractions as affected by moisture and fertiliser source and rate. Biology and Fertility of Soils. 2007;**44**(1):9-18

[59] Atiyeh RM, Subler S, Edwards CA, Bachman G, Metzger JD, Shuster W. Effects of vermicomposts and composts on plant growth in horticultural container media and soil. Pedobiologia (Jena) [Internet]. 2000;**44**(5):579-590. Available from: http://www.sciencedirect.com/science/article/pii/S0031405604700736

[60] Sifola MI, Barbieri G. Growth, yield and essential oil content of three cultivars of basil grown under different levels of nitrogen in the field. Scientia Horticulturae. 2006;**108**(4):408-413

[61] Reeve JR, Carpenter-Boggs L, Reganold JP, York AL, Brinton WF. Influence of biodynamic preparations on compost development and resultant compost extracts on wheat seedling growth. Bioresource Technology. 2010;**101**(14):5658-5666. DOI: 10.1016/j.biortech.2010.01.144

[62] Schönherr J. Calcium chloride penetrates plant cuticles via aqueous pores. Planta. 2000;**212**(1):112-118

[63] Kaya M, Atak M, Khawar KM. Çiftçi, Cemalettin Y, Özcan S. Effect of pre-sowing seed treatment with zinc and foliar spray of humic acids on yield of common bean (*Phaseolus vulgaris* L.). International Journal of Agriculture Biology. 2005;7(6):875-878. Available from: http://www.ijab.org

[64] De Swart EAM, Groenwold R, Kanne HJ, Stam P, Marcelis LFM, Voorrips RE. Non-destructive estimation of leaf area for different plant ages and accessions of *Capsicum annuum* L. Journal of Horticultural Science and Biotechnology. 2004;**79**(5):764-770

[65] Pant AP, Radovich TJ, Hue NV, Paull RE. Biochemical properties of compost tea associated with compost quality and effects on pak choi growth. Scientia Horticulturae. 2012;**148**:138-146

[66] Siddiqui Y, Islam TM, Naidu Y, Meon S. The conjunctive use of compost tea and inorganic fertiliser on the growth, yield and terpenoid content of *Centella asiatica* (L.) urban. Scientia Horticulturae. 2011;**130**(1):289-295. DOI: 10.1016/j.scienta.2011.05.043

[67] Tollefson SJ, Curlango-Rivera G, Huskey DA, Pew T, Giacomelli G, Hawes MC. Altered carbon delivery from roots: Rapid, sustained inhibition of border cell dispersal in response to compost water extracts. Plant and Soil. 2015;**389**(1-2):145-156

[68] Kim MJ, Shim CK, Kim YK, Hong SJ, Park JH, Han EJ, et al. Effect of aerated compost tea on the growth promotion of lettuce, soybean, and sweet corn in organic cultivation. The Plant Pathology Journal. 2015;**31**(3):259

[69] de Sanfilippo EC, Argüello JA, Abdala G, Orioli GA. Content of auxin-inhibitor-and gibberellin-like substances in humic acids. Biologia Plantarum. 1990;**32**(5):346-351

[70] Chen Y, Aviad T. Effects of humic substances on plant growth. In: MacCarthy P, Clapp C, Malcolm R, Bloom PR, editors. Humic Substances in Soil and Crop Sciences [Internet]. WI: SSSA; 1990. pp. 161-186. Available from: https://dl.sciencesocieties.org/publications/books/abstracts/acsesspublicati/humicsubstances/161

[71] Muscolo A, Bovalo F, Gionfriddo F, Nardi S. Earthworm humic matter produces auxin-like effects on Daucus carota cell growth and nitrate metabolism. Soil Biology & Biochemistry. 1999;**31**(9):1303-1311

[72] Valdrighi MM, Pera A, Agnolucci M, Frassinetti S, Lunardi D, Vallini G. Effects of compost-derived humic acids on vegetable biomass production and microbial growth within a plant (*Cichorium intybus*)-soil system: A comparative study. Agriculture, Ecosystems and Environment. 1996;**58**(2-3):133-144. Available from: https://www.sciencedirect.com/science/article/pii/0167880996010316

[73] Cacco G, Attinà E, Gelsomino A, Sidari M. Effect of nitrate and humic substances of different molecular size on kinetic parameters of nitrate uptake in wheat seedlings. Journal of Plant Nutrition and Soil Science. 2000;**163**(3):313-320. Available from: http://doi.wiley.com/10.1002/1522-2624%28200006%29163%3A3%3C313%3A%3AAID-JPLN313%3E3.0.CO%3B2-U

[74] Panuccio MR, Muscolo A, Nardi S. Effect of humic substances on nitrogen uptake and assimilation in two species of pinus. Journal of Plant Nutrition. 2001;**24**(4-5):693-704

[75] Auer CA, Motyka V, Březinová A, Kamínek M. Endogenous cytokinin accumulation and cytokinin oxidase activity during shoot organogenesis of Petunia hybrida. Physiologia Plantarum. 1999

[76] Spaccini R, Baiano S, Gigliotti G, Piccolo A. Molecular characterization of a compost and its water-soluble fractions. Journal of Agricultural and Food Chemistry. 2008

[77] Arancon NQ, Pant A, Radovich T, Hue NV, Potter JK, Converse CE. Seed germination and seedling growth of tomato and lettuce as affected by vermicompost water extracts (teas). HortScience. 2012;**47**(12):1722-1728

[78] Garcia Martinez I, Cruz Sosa F, LarqueSaavedra A, Soto HM. Extraction of auxin-like substances from compost. Crop Research Hisar. 2002;**24**(2):323-327. Available from: https://eurekamag.com/research/003/766/003766310.php

[79] Arancon QN, Edwards CA, Dick R, Dick L. Vermicompost tea production and plant growth impacts. Biocycle. 2001:51-52. Available from: www.biocycle.net

[80] Arancon NQ, Edwards CA, Bierman P. Influences of vermicomposts on field strawberries: Part 2. Effects on soil microbiological and chemical properties. Bioresource Technology. 2006;**97**(6):831-840. Available from: http://www.ncbi.nlm.nih.gov/pubmed/15979873

[81] Kale RD, Mallesh BC, Kubra B, Bagyaraj DJ. Influence of vermicompost application on the available macronutrients and selected microbial populations in a paddy

field. Soil Biology & Biochemistry. 1992;**24**(12):1317-1320. Available from: https://www.sciencedirect.com/science/article/abs/pii/003807179290111A

[82] Carpenter-Boggs L. Understanding the science. Diving into compost tea. Biocycle. 2005

[83] Panchakavya NK. A Manual. Goa, India: Mother India Press; 2003

[84] Bernal MP, Sánchez-Monedero MA, Paredes C, Roig A. Carbon mineralization from organic wastes at different composting stages during their incubation with soil. Ecosystems and Environment: Agriculture; 1998

Chapter 6

Efficacy of Different Substrates on Vermicompost Production: A Biochemical Analysis

Pawlin Vasanthi Joseph

Abstract

The rapid increase in the volume of waste is one aspect of the environment crisis, accompanying global development. Earthworms play an important role in the decomposition of organic matter and soil metabolism through feeding, fragmentation, aeration, turnover and dispersion. The type of substrates used and species of earthworms introduced plays a significant role in plant growth and yield. The waste to be stabilized should support an adequate biomass needed for effective processing. In the present study the vermicompost produced from banana as a substrate did not show a significant increase in NPK content from that of the control. On the other hand poultry waste and vegetable waste with goat dung showed significant increase in the NPK content. The enhancement of the vermicompost was probably due to mineralization of the organic matter containing proteins and conversion of ammonium nitrogen into nitrite. Mineralization and consequent mobilization of phosphorous by enhanced bacterial and phosphatase activities during vermicomposting leads to increase in Phosphorus. The earthworm processed waste materials contain high concentration of exchangeable potassium, due to enhanced microbial activity during the vermicomposting process, which accordingly enhanced the rate of mineralization. Vermicompost tends to hold more nutrients over larger periods without adverse effects on the environment.

Keywords: vermicompost, nitrogen, phosphorus, potassium, substrates, organic matter

1. Introduction

Solid waste is one of the growing problems in both developed and developing countries. Due to the rapid growth in industrialization, most of the rural populations have shifted towards the urban area in search of employment. The rapid increase in the volume of waste is one aspect of the environment crisis, accompanying global development.

Earthworms play an important role in the decomposition of organic matter and soil metabolism through feeding, fragmentation, aeration, turnover and dispersion [1]. Earthworms are involved in the recycling of nutrients, soil structure, soil productivity and agriculture, and their application in environment and organic waste management is well understood [2, 3]. They help in the degradation of substrate indirectly by affecting microbial population structure and dynamics and also directly since their gut is capable of undertaking cellulolytic activity. Thus products of cellulose hydrolysis are available as carbon and energy sources for other microbes that inhabit the environment in which cellulose is degraded and this availability forms the basis of many biological interactions.

There are about 3627 species of terrestrial earthworms in the world [4]. Sixty three species of earthworm from Sri Lanka of which 47 are considered as zoogeographically important to the Asian region have been recorded [5]. Vermiculture biotechnology promises to contribute in the 'second green revolution' by completely replacing the destructive agrochemicals which did more harm than good to both farmers and their farmland during the 'first green revolution' of the 1950–1960s.

Three major groups of earthworms based on ecological strategies have been recognized: the epigeics (Epiges), anecics (Aneciques) and endogeics (Endoges) [6]. Epigeic earthworms live in the soil surface and are litter feeders. Anecic earthworms are top soil species, which predominantly form vertical burrows in the soil, feeding on the leaf litter mixed with the soil. Endogeic earthworms preferably make horizontal burrows and consume more soil than epigeic and anecic species, deriving their nourishment from humus.

Vermicomposting is a mesophilic procedure, using microorganism and earthworms that are dynamic at 10–32°C. Vermiculture provide for the use of earthworms as a natural bioreactor for cost effective and eco-friendly waste management. Earthworm fecundity is based on the rate of cocoon production, hatching success of cocoons and number of offspring's emerging from each cocoons. The success of the composting depends upon the fecundity of the earthworm.

The type of substrate used and species of earthworms introduced plays a significant role in plant growth and yield. The waste to be stabilized should support an adequate biomass needed for effective processing. The time, cost and space requirements could compete economically with conventional methods of composting [7].

2. Substrates used for vermicomposting

2.1 Cow and goat dung

Vermicomposting of cattle and goat manure by *Perionyx excavatus* and their growth and reproduction performance was studied [8]. They concluded that cattle manure provided more nutritious and friendly environment to the earthworm than goat manure. The effects of Goat manure sludge, sewage and effective microorganisms on the composting of pine bark was studied [9]. The pine bark goat manure compost had more desirable nutritional properties than the pine bark and pine bark sewage sludge composts. It had neutral pH, C\N ratio and high amount of inorganic constituents.

2.2 Poultry waste

Poultry litter is the mix of bedding material, manure and feathers that result from intensive poultry production. This includes litter from meat chickens (broilers), egg laying chickens (layers) kept under barn conditions, turkeys, ducks and quails.

Limited available data presents numerous challenges while vermicomposting poultry litter. High ammonical nitrogen concentration, auto heating, and high bulk density are some of the major concerns that need to be addressed while vermicomposting poultry litter [10]. Poultry wastes contain significant amount of organic salts and ammonia that kill worms. So it is necessary to neutralize freshly deposited wastes by $CaCO_3$.

2.3 Fruit waste

The Indian state of Tamil Nadu is the largest producer of bananas in the country cultivating around 9 million metric tons (MT) annually, but inefficient postharvest practices lead to massive waste every year. An average of 30% or 2.7 million MT of Tamil Nadu's bananas currently goes to waste largely due to the absence of integrated cold chain infrastructure. Banana cultivation produces a huge amount of waste: approximately 30 tonnes of waste is generated per acre in one crop season from banana stem alone.

India produces around 2300 tonnes of papaya annually. In the past decade, the area under papaya cultivation in India has hugely increased following the introduction of Taiwanese and Hawaiian varieties. The processing operation of fruits and vegetables produce significant wastes as by-products, which constitute about 25–30% of a whole commodity group. The waste is composed mainly of seed, skin, rind, and pomace, containing good sources of potentially valuable bioactive compounds, such as carotenoids, polyphenols, dietary fibers, vitamins, enzymes, and oils, among others.

2.4 Paper waste

The Indian paper industry accounts for about 1.6 per cent of the world's production of paper and paper board. It is the 15th largest in the world and is one of the high priority industries having a bearing on the socio-economic development of the country.

India consumes almost 100 lakh tons of paper and paper boards. Paper Mills in the country are increasing their production and renovating their plants. By 2025, the demand for paper would increase to 2.5 crore metric tons. There is no effective collection mechanism for waste paper from offices and households. Newspapers are used for packaging. Muncipalities are not efficient in waste management network. There is lack of space for storage and sorting of waste paper. No proper co-ordination exists between the informal sector and the main supply chain of waste paper to paper industry (**Tables 1** and 2).

2.5 NPK analysis

In the present study different substrates have been used to culture earthworms and the nutrient content of the vermicompost produced by them has been analysed. The nitrogen content has significantly increased in papaya waste, paper waste, poultry litter and vegetable waste with goat dung. Phosphorus content has significantly increased in all the wastes except banana and levels of potassium have decreased in banana and paper waste. In the study the vermicompost produced from banana as a substrate did not show a significant increase in NPK content from that of the control. On the other hand poultry waste and vegetable waste with goat dung showed significant increase in the NPK content.

In vermicompost, a higher amount of organic carbon is used when compared to the normal compost as the earthworms have higher additional assimilating capacity besides microorganisms. Earthworms also modify the conditions which subsequently lead to increased carbon losses as CO_2 due to microbial respiration in organic matter being converted to vermicompost [11].

Samples	Banana waste			Papaya waste			Paper waste		
	N	P	K	N	P	K	N	P	K
Control 45 days	0.48 ± 0.03*	1170 ± 5.83*	23 ± 0.37*	0.42 ± 0.06*	787 ± 6.12*	21 ± 0.32*	0.53 ± 0.30*	211.23 ± 4.38*	628.50 ± 93.04^{NS}
Treated 45 days	0.41 ± 0.07*	840 ± 2.55*	17.1 ± 0.32*	0.47 ± 0.05*	974 ± 9.08*	19.10 ± 0.28*	0.57 ± 0.09*	270.13 ± 21.92*	526 ± 149.66^{NS}

*Values are mean ± SD. *- $p \leq 0.05$; NS – Not significant. Refs. [30, 31]*

Table 1.
NPK analysis of vermicompost produced from banana waste, papaya waste and paper waste.

Samples	Poultry waste			Cow dung and vegetable waste			Goat dung and vegetable waste		
	N	P	K	N	P	K	N	P	K
Control 45 days	1.08 ± 0.07^{NS}	0.39 ± 0.05*	0.22 ± 0.02*	6.31 ± 469.077^{NS}	846 ± 48.256*	16.89 ± 127.71^{NS}	6.31 ± 469.077^{NS}	846 ± 48.256*	16.89 ± 127.71^{NS}
Treated 45 days	2.5 ± 0.1^{NS}	0.67 ± 0.14^{NS}	0.65 ± 9.11^{NS}	6.14 ± 434.925^{NS}	896 ± 40.329*	17.74 ± 80.233^{NS}	6.84 ± 1104.70^{NS}	1024 ± 25.292*	18.76 ± 83.66^{NS}

*Values are mean ± SD. *- $p \leq 0.05$; NS – Not significant.*

Table 2.
NPK analysis of vermicompost produced from poultry waste, cow dung and vegetable waste and goat dung and vegetable waste.

The pH reduction may be due to the mineralization of nitrogen into nitrates/nitrites and phosphrous into orthophosphates as well as bioconversion of organic wastes into organic acids [12]. Studies where *Bacillus* has been reported to be isolated from the gut of *Eisenia foetida* [13] and these gut associated miroflora assists the earthworms significantly to hasten the decomposition of organic matter by producing certain enzymes namely cellulase, amylase, protease etc. Although dependent upon earthworm species, it is known that earthworms interact with microorganisms (fungi, bacteria and actinomycetes) on three broad spatial scales-burrow linings, casts and earthworm gut or intestine. Importantly, the increased gut associated microflora are then excreted throughout the media within earthworm casts and via microbial adherence to earthworm skin whilst the transit and dispersal mechanisms associated with the water flow also help to further dissipate microorganisms [14].

2.6 Nitrogen

The enhancement of the vermicompost was probably due to mineralization of the organic matter containing proteins [15, 16] and conversion of ammonium nitrogen into nitrite [17, 18]. The final N content of the compost as well as the vermicompost depends on the initial content of N in the substrate and the extent of its decomposition [19, 20]. The earthworms can enhance N levels during vermicomposting through the digestion of substrate in their gut and simultaneous addition of nitrogenous excretory products, mucous, body fluid, enzymes; besides the decay of dead tissues of worms in vermicomposting system [21]. This nitrogen content value could have been due to the nitrogenous metabolic products of earthworms which are returned to the vermicompost as casts.

2.7 Total phosphorous

Mineralization and consequent mobilization of phosphorous by enhanced bacterial and phosphatase activities during vermicomposting leads to increase in P [22]. An increase of 25% P in paper waste sludge after the activities of earthworms was reported [23]. They further suggested that the consequent increase in P after the earthworm's activities may be due to the direct action of worm gut enzymes and due to enhanced microbial activity in the vermicompost. Increase in P content in vermicompost could be due to enhanced mineralization and mobilization of phosphorous as a result of increased bacterial and fecal phosphatase activity of earthworms [22].

Plant litter was found to contain more available P after ingestion by earthworms, which may be due to the physical breakdown of the plant materials by worms. An increase of 25% in P in paper waste sludge after worm activity was observed. They attributed this increase in P to the direct action of worm gut enzymes and indirectly by stimulation of the microflora [23].

The increased phosphorous level was due to mineralization of phosphorous. The release of phosphorous in the available form is performed partly by earthworm gut phosphatases and further release of phosphorous might be assigned to the phosphorous solubilizing microbes present in vermicast. The earthworm affects phosphorous mineralization in wastes during passing organic matter through its gut.

2.8 Total potassium

Decrease in potassium content in the vermicompost may be due to the leaching of this soluble element through the action of excess water draining through the mass [24].

The rate of nutrient loss was directly related to the initial concentrations [25]. The selective feeding of earthworms on organically rich substances which breakdown during the passage through the gut, biological grinding, together with enzymatic influence on finer soil particles, were lightly responsible for increasing the different forms of K [26]. The increase of soil organic matter resulted in decrease K fixation and subsequent increase K availability [27].

The available micro-nutrients like potassium (K) are required for assimilation by earthworms during the vermicomposting, although the quantity required is very low as compared to the initial content present in the parent feed material. The production of acids by the microorganisms and enhanced mineralization rate through increased microbial activity during the vermicomposting process play a key role in the solubilizing of insoluble potassium [28, 29].

The increase of potassium in the treated might be due to changes in the distribution of potassium between exchangeable and non-exchangeable forms. The earthworm processed waste materials contain high concentration of exchangeable potassium, due to enhanced microbial activity during the vermicomposting process, which accordingly enhanced the rate of mineralization.

When organic matter passes through the gut of earthworm, unavailable potassium is transformed to more soluble forms with enhanced rate of mineralization. Decomposition of organic material by microorganisms produces acid products that increase the available soluble potassium. On the other hand, the gut of earthworm has a big population of microflora that could enhance potassium content in the vermicompost.

3. Conclusion

Vermicomposting has many applications such as increasing water holding capacity, crop growth and yield, improves the physical, chemical and biological properties of the soil. It increases the production of plant growth regulators. Vermicompost is pollution free and cost effective. The texture of vermicompost is homogenous, contains many plant growth hormones and soil enzymes and tends to hold more nutrients over larger periods without adverse effects on the environment.

Acknowledgements

The author would like to thank the Department of Biotechnology, Government of India, for strengthening Life Sciences and for initiating Research Projects at the under graduate level.

Conflicts of interest

There is no conflict of interest.

Author details

Pawlin Vasanthi Joseph
Nirmala College for Women, Coimbatore, Tamilnadu, India

*Address all correspondence to: pawl_06@rediffmail.com

References

[1] Singh D, Suthar S. Vermicomposting of herbal pharmaceutical industry solid wastes. Ecological Engineering. 2012;**39**: 1-6

[2] Ansari AA, Ismail SA. Role of earthworm in vermitechnology. Journal of Agricultural Technology. 2008;**8**(2): 403-415

[3] Ansari AA, Sukhraj K. Effect of vermiwash and vermicompost on soil parameters and productivity of okra (*Abelmoschus esculentus*) in Guyana. Journal of Agriculture and Environmental Sciences. 2010;**8**(6):666-671

[4] Reynolds J. Earthworms of the world. Global Biodiversity. 1994;**4**:11-16

[5] Stephenson J. The Fauna of British India Including Ceylon and Burma. London: Taylor and Francis; 1923. p. 518

[6] Bouché MB. Relations entre les structures spatiales et fonctionelles des écosystemes, illustrées par le rôle pédobiologique des vers de terre. In: Pesson P, editor. La Vie dans les Sols, Aspects Nouveaux, Études Experimentales. Paris: Gauthier-Villars; 1971. pp. 187-209

[7] Haimi J, Huhta V. Capacity of various residues to support adequate earthworm biomass for vermicomosting. Biology and Fertility of Soils. 1986;**2**:23-27

[8] Loh TC, Lee YC, Linang JBT. Vermicomposting of cattle and goat manure by *Eisenia fetida* and their growth reproduction performance. Bioresource Technology. 2005;**96**(1): 111-114

[9] Mupondi LT, Mnkeni PNS, Brutsch MO. The Effects of Goat Manure, Sewage Sludge and Effective Microorganisms on the Composting of Pine Bark. Compost Science and Utilization. 2006;**14**(3):201-210

[10] Natarajan N, Gajendran M. Vermiconversion of paper mill sludge for recycling the nutrients using earthworm *Eudrilus eugeniae*. International organization of scientific research. Journal of Environmental Science, Toxicology and Food Technology. 2014;**8**(9):06

[11] Aira M, Monroy F, Dominguez J. *Eisenia fetida* (Oligochaeta: Lumbricidae) modifies the structure and physiological capabilities of microbial communities improving carbon mineralization during vermicomposting of pig manure. Microbial Ecology. 2007a;**54**:662-671

[12] Ndegwa PM, Thompson SA, Das KC. Effects of stocking density and feeding rate on vermicomposting of biosolids. Bioresource Technology. 2000;**71**:5-12

[13] Jyotsana P, Vijayalakshmi K, Prasanna ND, Shaheen SK. Isolation and characterization of cellulase producing *Lysinibacillus sphaericus* from gut of *Eisenia foetida*. The Bioscan. 2010;**6**(2): 325-327

[14] Edwards CA, Bohlen PJ. Biology and Ecology of Earthworms. London: Chapman and Hall; 1996. p. 456

[15] Bansal S, Kapoor KK. Vermicomposting of crop residues and cattle dung with *Eisenia fetida*. Bioresource Technology. 2000;**73**:95-98

[16] Kaushik P, Garg VK. Vermicomposting of mixed solid textile mill sludge and cow dung with the epigeic earthworm *Eisenia foetida*. Bioresource Technology. 2003;**90**(3):311-316

[17] Suthar S, Singh S. Feasibility of vermicomposting in biostabilization of sludge from distillary industry. Science of the Total Environment. 2008;**394**: 237-243

[18] Atiyeh RM, Dominguez J, Subler S, Edwards C. Changes in biochemical properties of cow manure during processing by earthworms (*Eisenia andrei bouche*) and the effects on seedling growth. Pedobiologia. 2000;**44**: 709-724

[19] Crawford JH. Review of composting. Process Biochemistry. 1983;**18**:14-15

[20] Gaur AC, Singh G. Recycling of rural and urban waste through environmental and vermicomposting. In: Tandan HLS, editor. Recycling of Crop, Animal, Human and Industrial Waste in Agriculture, Fertilizer. New Delhi: Development and consultation organization; 1995. pp. 31-44

[21] Suthar S. Vermicomposting potential of *Perionyx Sansibaricus* (Perrier) In different waste materials. Bioresource Technology. 2007;**98**: 1231-1237

[22] Edwards CA, Lofty JR. Biology of Earthworms. London: Chapman and Hall; 1972

[23] Satchel JE, Martin K. Phosphatase activity in earthworm faeces. Soil Biology and Biochemistry. 1984;**16**: 191-194

[24] Tahir TA, Hamid FS. Vermicomposting of two types of coconut wastes employing *Eudrillus eugeniae*: A comparative study. International Journal of Recycling of Organic Waste in Agriculture. 2010;**1**(7):1-6

[25] Das D, Powell M, Bhattacharya P, Banik P. Changes of carbon, nitrogen, phosphorous and potassium content during storage of vermicomposts prepared from different substrates. Environmental Monitoring and Assessment. 2014;**186**:8827

[26] Rao S, Subba Reo A, Takkar PN. Changes in different forms of K under earthworm activity. In: National Seminar on Organic Farming and Sustainable Agriculture; India: 1996; pp. 9-11

[27] Olk DC, Cassman KG. Reduction of potassium fixation by organic matter in vermiculitis soils. In: Soil Organic Matter Dynamics and Sustainability of Tropical Agriculture. (In Agris). USA: John Wiley and sons Ltd; 1993. pp. 307-315

[28] Kaviraj, Sharma S. Municipal solid waste management through vermicomposting employing exotic and local species of earthworms. Bioresource Technology. 2003;**90**:169-173

[29] Khwairakpam M, Bhargava R. Vermitechnology for sewage sludge recycling. Journal of Hazardous Materials. 2009;**161**(2–3):948-954

[30] Joseph PV. Microbiological and physicochemical analysis of vermicompost of fruit waste by *Eudrilus eugeniae*. International Journal of Science and Research Methodology. 2018;**8**(4):77-93

[31] Joseph PV. NPK ratio of paper waste degradation through vermicomposting in an Institutional setup. In: National Seminar on Eco-Waste Management NS Nanobiology; 2016

Chapter 7

Organic Fertilizer Production and Application in Vietnam

Pham Van Toan, Ngo Duc Minh and Dao Van Thong

Abstract

Crop production is an important subsector of Vietnam's agriculture, has an impressive achievement in last 30 years and based on the intensive production with increasing use of chemical fertilizer and pesticide. Consequences are the negative effects on environment and human health and food safety. Organic agriculture has become a trend worldwide and is developing rapidly in the world. In Vietnam the certified organic farming area has expanded since 2012. Organic market revenue in Vietnam is estimated to be at $132.15 million a year. Most Vietnamese certified organic products are exported to international markets. Organic agriculture using organic fertilizer is one of Vietnam government's priorities. Vietnam already produced organic fertilizer from different materials by using different production technologies, but the production capacity is small and does not meet the demand for organic agriculture. Vietnam government encourages, promotes the organic fertilizer production, application and has the policy to develop the organic fertilizer in Vietnam.

Keywords: Vietnam agriculture, organic agriculture, organic fertilizer

1. Introduction

Vietnam is one of the most biodiversity countries with 13,200 terrestrial plant species, around 10,000 animal species and 3000 aquatic species. The country also has an extremely long coastline extending over 3260 km, but Vietnam is the country most vulnerable to climate change and frequent natural disasters in Southeast Asia. Agriculture is the most important economic sector in the country and more than 70% of Vietnam's population is dependent on it. In the period 2000–2018, the output value of agriculture, forestry and fisheries continued to increase with the average rate of more than 4%/year. In terms of value-added of agriculture, the average growth rate of 3.7%/year of GDP in that period is relatively high and stable. The structure of agricultural production has gradually shifted to the higher efficient sector which is associated with market demand. Agricultural production has gradually improved to meet domestic needs. Despite market fluctuations, natural disasters, complicated epidemics, food production continues to grow in absolute value. Agriculture, forestry and fisheries are the only sectors of Vietnam to have consecutive trade surplus, even in the phase of difficult economic state. It shows the evident comparative advantages of Vietnam's agriculture demonstrating the important role of agriculture in the balance of payments of Vietnam's economy.

Crop production plays a very important role in Vietnam's agriculture. After more than 30 years of renovation, the crop production subsector has made an important contribution to bringing Vietnam from a food shortage and importing food country to become one of world leading agricultural exporters. The economic value of the crop production sub-sector currently contributes over 70% of the agricultural sector's GDP and nearly 50% of the agricultural-forestry-fishery export value, contributing to hunger elimination, poverty reduction, food and social security. The crop production is now continuing to develop towards commodity production, high quality, sustainable production, climate change adaptation and export-oriented. According the report of ministry of agriculture and rural development in 2018, the export turnover of agricultural products in crop production sub-sector reached 18.9 billion USD of the total 40 billion USD of agricultural sector exporting value. Among the 10 major export commodities (over 1 billion USD) of the whole sector, there are seven commodities from crop production as rice, coffee, cashew, fruits, vegetables, rubber, cassava and pepper. Export results achieved in 2018 affirmed Vietnam's position as an exporter of agricultural products, ranked fifteenth in the world in export value and exported to 180 countries and territories around the world [1].

Crop production growth in Vietnam is based on intensive natural resource, increasing use of fertilizers, plant protection chemicals. While achieving economic targets, agricultural production causes adverse environmental effects, imbalance and depletion of natural resources. Weaknesses in the management of water resources and agricultural residues also cause increasing pollution and greenhouse gas emissions. Pollution has started to impact on soil fertility and yields, the effectiveness of chemicals in combating pests and disease, farmer health, environmental health and the safety of food. Meanwhile, the wasteful use of inputs is a drag on farm profitability. Though the incidence and impacts of agricultural pollution in Vietnam remains limited, but more has started emerging. Meanwhile, the Vietnamese has become increasingly aware of the human and environmental health problems that agricultural pollution is generating. Organic production used organic fertilizer and is one of target goal of sustainable development of crop production in Vietnam.

2. Crop production and fertilizer, pesticide consumption in Vietnam

2.1 Crop production achievement

In 2016 the total crop production area in Vietnam are 11,527,000 ha, in which rice area is 4,136,000 ha and others annual crop planting area is 2,852,000 ha. Perennial crop cultivating area is 4,539,000 ha includes the key commodity crops like rubber, coffee, cashew, pepper, tea and fruit trees [2].

In general, the yields of major crops are stable in last 5 years. The average yield of major food crops is about 5.5–5.8 tons/ha for rice, 4.4–4.8 tons/ha for maize (**Figure 1**) and industrial crops is 19–19.5 tons/ha for cassava, 2.4–2.5 tons/ha for coffee, 2.2–2.5 for rubber, 0.7–0.8 for cashew (**Figure 2**). In 2018, the vegetable and fruit production in Vietnam grow rapidly and reach the exporting value of 3.8 billion USD and increase 9.2% to year 2017. Vietnam is the biggest pepper exporter in the world with the amount of more than 200,000 tons/year. The average yield of Vietnamese pepper is 2.2–2.5 tons/ha and 2.6-fold higher as compared to average yield of pepper all over the world.

Despite some objective difficulties, key agricultural products (coffee and cashew) still maintained high export values. Export results achieved in 2018 (**Figure 3**) affirmed Vietnam's position as an exporter of agricultural products, ranked fifteenth in the world in export value and exported to 180 countries and territories around the world [1].

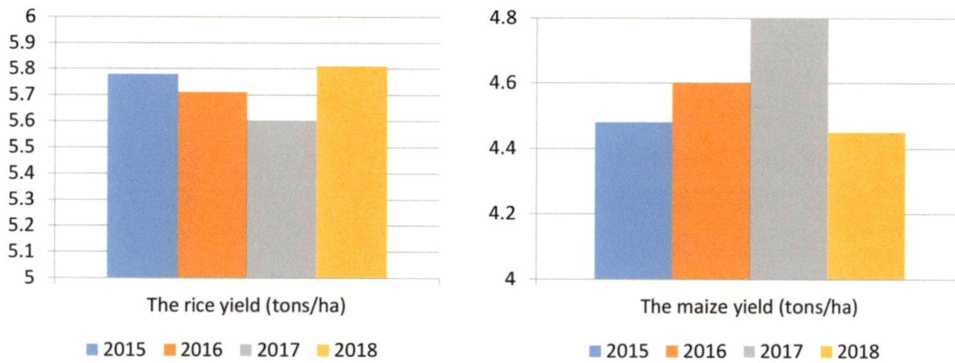

Figure 1.
The yield of some food crops in Vietnam.

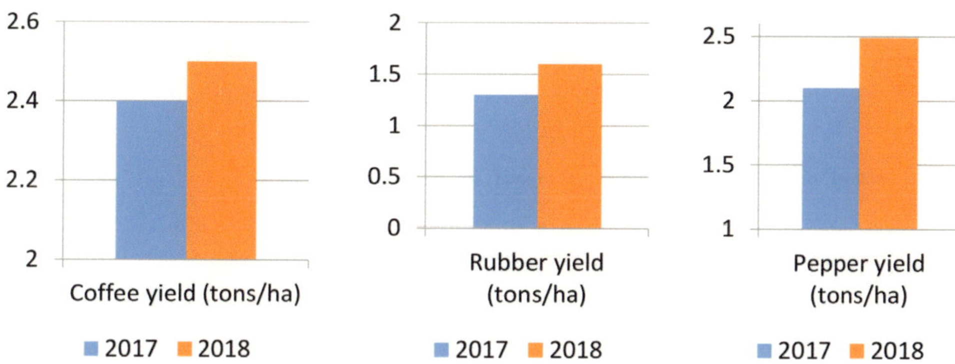

Figure 2.
The yield of some major industrial crops.

2.2 Fertilizer and pesticide consumption

Together with the trend of agricultural intensification, the use of inputs, especially fertilizers and pesticides for crops, also increased very quickly in the past two decades. The country has imported between 3.5 million and 4.5 million tons inorganic fertilizers per year since 2000. Urea imports reached a peak in 2000–2004, before declining but amounts of imported ammonium sulfate and potassium have tended to increase since. From 1985 to 2005, the rate of fertilizer consumption of nitrogen, phosphorous, and potassium increased by about 10% per year, peaking at 25 million tons in 2005. Fertilizer use for crops has varied among and within provinces, but generally increased in volume over time. Fertilizer application rates vary greatly, depending on the types of crops, varieties, cropping seasons, locations, soil types, and forms of application. Overall, fertilizer use in crop cultivation has been increasing. In general, crop requiring the most fertilizer application is rice, accounts for approximately 65% of total fertilizer demand, followed by corn crop with 9%. Short duration growing crop such as sugarcane, peanuts, soybeans, cotton, vegetables etc. use 6% of fertilizer; the other plants including rubber, coffee, tea, pepper, cashew, fruit, etc. account for 20% (**Figure 4**).

There are three main cropping seasons in Vietnam: Winter-Spring from late November to March of the following year, Summer-Autumn from April to August and Autumn-Winter from late August to late November. Agricultural production mainly concentrates in the Winter-Spring season. The demand in Winter-Spring crop accounts for 49% of total fertilizer demand per year, the other two seasons have relatively equal demand of about 25% of total demand. Fertilizer demand in

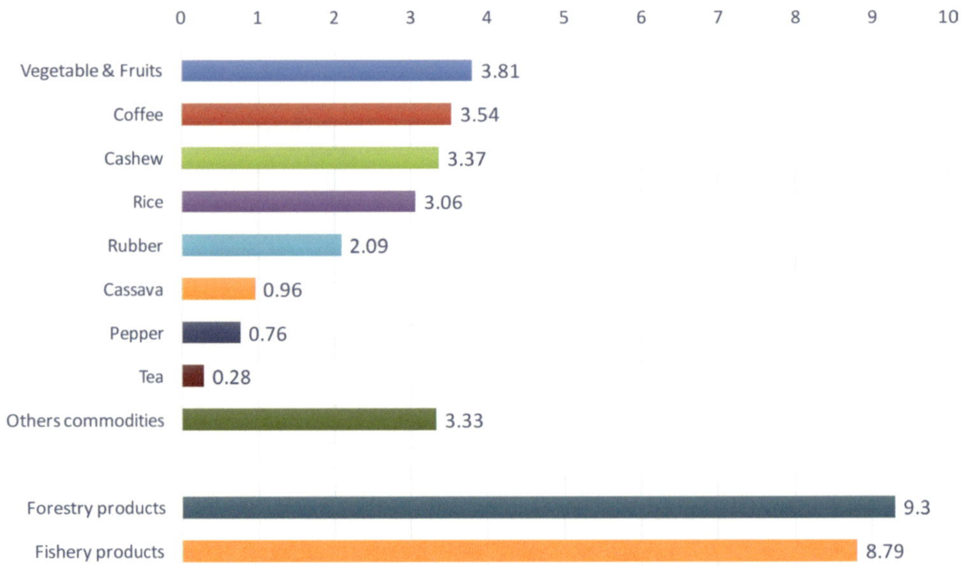

Figure 3.
Export value of some agricultural commodities in 2018 (source: Vietnam Customs [3]).

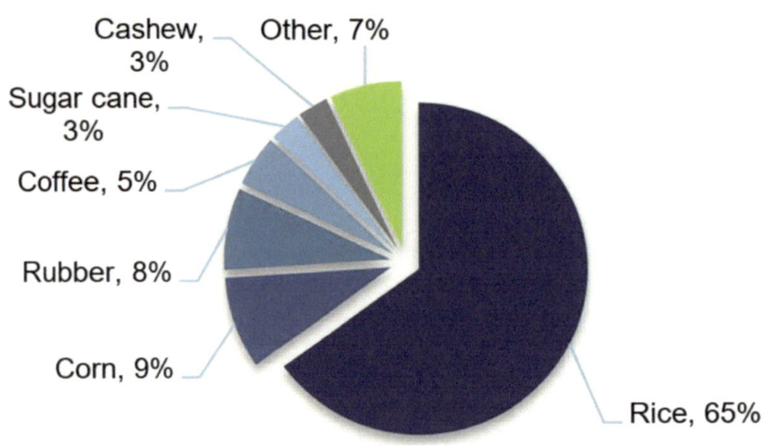

Figure 4.
The fertilizer use by crop in Vietnam (source: Tin [4]).

Northern Vietnam is clearly separated by different time of a year. While Summer-Autumn season takes up only 6% of total demand that of Winter-Spring season is up to 58% of fertilizer demand. Fertilizer demand in Central Vietnam and Southern Vietnam don't have that difference and f is relatively stable throughout the year. Briefly, Winter-Spring has seasonally highest fertilizer demand (**Figure 5**) [4].

Annually, Vietnamese farmers spend about VND 110.000 billion (about USD 5 billion) on fertilizers. Compared with nearby countries, Vietnam fertilizer consumption is only lower than China in terms of fertilizer use dosage. Vietnam farmer apply NPK fertilizer of dosage 297 kg/ha. The Vietnam fertilizer market was estimated at USD 228.1 million in 2017 and is expected to reach to USD 280.9 million by 2023, growing at a CAGR of 4%. Currently, the market is less regulated, less technologically, highly competitive, and has good opportunities for growth [5].

Similar to fertilizers, the consumption of pesticides in Vietnam has increased dramatically in the 2 past decades together with the intensification of the agricultural

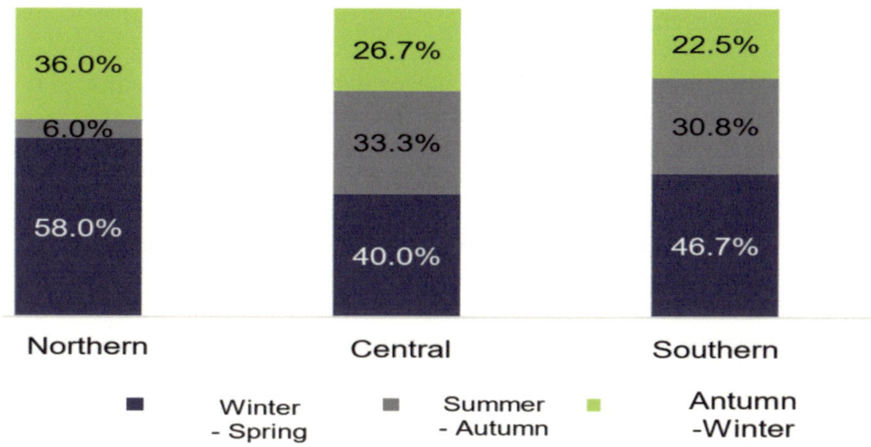

Figure 5.
Fertilizer demand by season and region (source: Tin [4]).

sector. In 1981–1986, Vietnam imported around 6500–9000 tons of pesticide active ingredients, then increased to 13,000–15,000 tons/year in 1986–1990, to 20,000–30,000 tons/year in 1991–2000, to 33,000–75,000 tons/year in 2001–2010 and up to approximately 100,000 tons/year around 2015 [6]. Along with that trend, the import value of pesticides increased quickly from around US $472 million in 2008 to US $537 million in 2010 and nearly US $700 million in recent years [2]. In 10 years (2000–2011), the number of pesticides registered and used in Vietnam has increased 10 times. In 10 years (2000–2011), the number of pesticides registered and used in Vietnam has increased 10 times. Before 2000, the number of active ingredients was around 77, corresponding to 96 trading products and increased in 2011 up to 1202, corresponding to 3108 trading products [7].

Vietnam, as previously noted, has achieved high rates of growth in agricultural output over the past decades, but this accomplishment has been at a heavy cost to the environment. The sector's increasing use of land and synthetic inputs has accelerated deforestation, biodiversity loss, land degradation, water pollution, and greenhouse gas emissions. Saddled as it is with high expectations, Vietnamese agriculture will need to produce more from less going forward. Maintaining high output growth under changing climate and economic conditions may require a strategy of intensification, sparing not only time and labor, but also land and water, pesticides and fertilizer.

Recently, the Vietnamese government has policies to shift away from production to focus more on quality, value addition and sustainability. This strategic shift was highlighted in Decision no. 899/QD-TTg dated 10th June, 2013 on approving the plan of restructuring the agricultural sector towards improving added value and sustainable development. The agricultural restructuring plan (ARP) defines sector goals in terms of the triple bottom line of economically, socially, and environmentally sustainable development. It lays out expected changes in the roles and spending patterns of the government in the sector and discusses the need to work with other stakeholders, including in the private sector. There are currently many initiatives aiming in these directions. Yet achieving the shift these represent on a large, sector-wide scale, will require important changes in certain economy-wide and sector-specific policies and, over time, major changes and additions to the core institutions servicing agriculture. It calls for an ambitious and ongoing process of learning and experimentation, and several potential directions are offered below for consideration. Various programs have been initiated in Vietnam to promote sustainable production and natural resource management practices.

3. Vietnam organic agriculture

Organic agriculture has become a trend worldwide. Organic agriculture is developing rapidly in the world with 57.8 million ha and the market potential worth nearly US $90 billion [8]. In Vietnam, organic agriculture journey ultimately led to establishment of The Vietnam Organic Agriculture Association (VOAA) within the first congress of organic agriculture development held in Hanoi in May 2012. From these important steps, the certified organic farming area has expanded during last 5 years in Vietnam. According to the Research Institute of Organic Agriculture (FiBL) and the International Federation of Organic Agriculture Movements [8], the certified area of organic agricultural production in Vietnam increased rapidly, from 43,000 ha in 2014 to 118,000 ha in 2016 [9].

Up to now, 33 of the 63 provinces and cities nationwide have developed organic farming and aquaculture models. About 60 groups, corporations and production establishments have invested in organic agriculture in Vietnam. Though organic farming area is modest as compared to the total farming area in Vietnam, businesses and organizations are applying international organic standards and certified organic products are being exported to many markets, including the US and EU. Organic market revenue in Vietnam is now estimated to be at $132.15 million a year, with spending for such products in the north higher than that in the south. Most Vietnamese certified organic products are exported to international markets such as Taiwan, Singapore, Japan, EU countries, the United States and Australia. Nearly 80 domestic companies have been certified by the EU.

In 2015, the Ministry of Science and Technology (MOIT) issued TCVN 11041: 2015 to guide the production, processing, labeling and marketing of food produced by organic methods. Within international collaborative projects or by private and/or foreign enterprises that based on different standards such as: The Participatory Guarantee System (for organic vegetable); EU, USDA, JAS standards (for organic tea, rice, vegetables, fruits) most of Vietnamese organic agricultural products are based on the foreign standards but not according to the TCVN 11041: 2015. In 2018, Ministry of Science and Technology (MOIT) has officially issued the first standards system for The National Organic Agriculture Standards (production, cultivation, animal husbandry, processing and labelling of organic products) putting an end to any argument related to actual criteria of organic agriculture, as well as responding to expectations of farmers and enterprises in this field. With referred to IFOAM's standards and standards of several countries with advanced organic agriculture including the U.S., EU, Japan, Thailand, and China, the new the Vietnamese Organic Agriculture Standards is in line with the current standards adopted by the ASEAN countries under the ASEAN Standard for Organic Agriculture. This is the important reference for farmers and producer to practice organic agriculture and for appropriate authorities to inspect, control the organic agricultural production. The standard is supposed to promote agricultural production in general and organic agriculture in particular, helps add more values to products improve quality of domestic and export goods.

In 2018, the government issued Degree 109/2018/ND-CP providing preferential terms for small enterprises, cooperatives, farms and farmer households engaged in organic agriculture. According to the decree, the government will fund all organic product certification costs and cost of verifying areas eligible for organic production. Farmers and cooperatives can also enjoy the government's agricultural promotion assistance in organic production training. This decree which takes effect on October 15, is actually an important legal framework for organic farming, and on that basis, mobilizes all economic sectors, enterprises, cooperatives. The Government of Vietnam always strongly supports efforts to develop a sustainable and environmentally friendly agriculture, improving the productivity and competitiveness of products, including organic agricultural products. In recent years, Vietnam has tried to complete national organic standards, comprehensive legal framework for production, certification and quality control of organic agricultural products and support policies to promote organic agriculture development.

4. Organic fertilizer production in Vietnam

In Vietnam organic fertilizers are fertilizers produced from the main raw materials that are natural organic substances (excluding synthetic organic substances), processed through physical or biological methods. Organic fertilizer composited mainly of organic substances and nutrients derived from organic materials (Degree No 108/ND-CP). In combination with mineral nutrition elements or beneficial microorganisms, organic fertilizer can be called as organic mineral fertilizer or biological-organic fertilizer or bio-organic fertilizer. According Degree No 108/ND-CP, organic fertilizer should be free of Salmonella while the density of *E. coli* and Coliform is lower than 1.1×10^3 MPN/g. The heavy metal concentration in organic fertilizer do not exceed 10.0 ppm for as, 5.0 ppm for Cd, 200.0 ppm for Pb and 2.0 ppm for Hg. The main quality requirement of organic fertilizer is showed in **Table 1**.

4.1 Organic fertilizer production technology

The organic fertilizer production line is commonly used to process different fermented organic substance into biological organic fertilizer. In Vietnam, organic fertilizers are now produced domestically in two ways: traditional composting and industrial production.

Traditional composting methods are mainly used on farm scale based on waste materials or crop residues collected from livestock and household farming. The traditional composting procedures take as long as 4–8 months to produce finished compost, by which organic residues are mixed well, and mineral elements can be added, and then composted into piles.

The industrial production of organic fertilizer is production of compost in industrial scale by using the rapid composting technology. Rapid composting methods offer possibilities for reducing the processing period up to some weeks. The industrial organic fertilizer production needs to invest in infrastructure, equipment lines with the large production capacity. Currently, there are in Vietnam 180 enterprises granted licenses to produce organic fertilizers, accounting for 24.5% of the total production licenses granted by authority agency with the production facilities about 2.5 million tons/year, accounting for 8.5% of the total capacity of domestic fertilizer production [10].

In Vietnam, rapid composting methods are used in early 2000 based on guide line of FAO on the on farm composting methods [11], in which the compost is regulated at 10% of oxygen saturation, the moisture from 50 to 55%, the pH from 5.5 to 7.0, the carbon-nitrogen (C:N) ratio around 20:1, the size of the parent materials from 5 to 10 cm. As compost enrichment the biomass of decomposing bacteria, fungi added at the rate of 500–1000 g/tons of compost materials [12].

There are several composting technologies in Vietnam as following: pile composting, box chamber composting, open-furrow composting with turning and aeration and enclosed vessel composting with mechanical agitation and aeration. Generally, less capital investments in equipment mean less capacity to treat wasted organic materials. These materials also need longer composting periods to reach

Kind of organic fertilizer	Quality parameters	Measured unit	Standards
Traditional organic fertilizer	Organic matter (OM)	%	≥20.0
	C/N		≤12.0
	Moisture	%	≤30.0
	pH H_2O		≥5.0
Bio-organic fertilizer	Organic matter (OM)	%	≥15.0
	Density of beneficial microbes or	CFU/g	$\geq 1.0 \times 10^6$
	Number of infective propagules of mycorhiza	IP/g	≥10
	Moisture	%	≤30.0
	pH H_2O		≥5.0
Biological organic fertilizer	Organic matter (OM)	%	≥20.0
	Humix acid, fulvic acid or	% of OC or	≥2.0
		%	≥3.5
	Other biological substances	According the standards or regulation	
	Moisture	%	≤30.0
	pH H_2O		≥5.0
Organic mineral fertilizer	Organic matter (OM)	%	≥15.0
	Content of total nitrogen available phosphorus and potassium	%	≥8.0 ≤ 18.0
	Content of each total nitrogen, available phosphorus and potassium	%	≥2.0
	Moisture	%	≤25.0
	pH H_2O		≥5.0

Table 1.
Quality requirements of organic fertilizer in Vietnam.

maturity. In contrast, more capital investments in equipment mean more capacity and efficiency for composting organic materials. In general, domestic organic fertilizer production facilities now invest in simpler production technologies. Basic organic fertilizer production line equipment including excavators; turning machine; crusher and screen; drying system; additive pumping system, microbial spray; weighing and packaging system of finished products. Most equipment lines are created in the country. Some organic fertilizer production facilities from waste, livestock waste, and crop residues have invested in the installation of advanced equipment lines from developed countries like Germany, Belgium, Netherlands, and Japan. Advanced production technologies allow to shorten the composting processing time by precisely adjusting the composting temperature, moisture, pH combined with the use of cellulolytic microorganism to create high quality organic fertilizer products. In addition to the mechanization and automation of the process of collecting, treating, supplying, crushing and sifting materials; the process of drying, granulating and bagging in modern production lines allows increasing labor productivity, production capacity and reducing production costs.

4.2 Materials for organic fertilizer production

The raw material of organic fertilizer can be used as agricultural waste, animal waste, industrial waste, household waste, municipal sludge and peat after safety disposal and fermentation, these materials are made into organic fertilizer. Thus, organic fertilizer contains a variety of organic acids, peptides, and rich nutrients including nitrogen, phosphorus and potassium. Not only provide comprehensive nutrition for crops, also with long fertilizer effect, which can Increase and update the soil organic matter and promote microbial breeding, improve soil physical and chemical properties and biological activity [13]. The sources of materials for organic fertilizer production in Vietnam are now diverse and abundant, including waste from animal husbandry, aquaculture, agricultural product processing, crop residues, peat, and domestic waste. Microbial inoculants, mineral elements, biological supplements to improve the quality and efficiency of fertilizer can be used [14].

According Vietnam General Statistic Office in 2015, Vietnam produced 45.22 million tons of rice, 5.28 million tons of maize, 10.67 million tons of cassava, 1.445 million tons of coffee and 18.320 million tons of sugar cane. Based on crop biomass and product, (Trinh) [15] calculated the agricultural waste approximate 76 million tons including 45.22 million tons of rice straw, 8.73 million tons of rice husk, 4.04 million tons of sugarcane bagasse (SCB), 6.33 million tons of maize by-products, 1 million tons of coffee shell and 10 million tons of vegetable by-products [15]. Agricultural waste contained not only the carbohydrate composition and plant essential nutrition like NPK and microelement [14] (**Table 2**).

As of April 2017, Vietnam has 2,519,411 buffaloes; 5,496,557 cows; 28,312,083 pigs and 341,892,000 poultry and estimated to release about 85 million tons of solid waste [1]. Animal waste has organic content; elements plurality of minerals is quite high and contains almost medium micro nutrients which help soil fertility [14], **Table 3**.

Vietnam exported every year more than 7 million tons of seafood products and made more than 5 million tons of seafood by-products that can be used as raw material for organic fertilizer production [15]. Seafood processing by product is protein, lipid and micro element (**Table 4**).

At present Vietnam has no standard for raw materials of organic fertilizers in regulations regarding fertilizer production, distribution, and use [16]. Varied raw materials and poorly controlled manufacturing could cause a wider range of nutrient content of domestic "organic fertilizers" compared with that of the imported ones.

According Hien [14], Vietnam has about 7.1 billion cubic meters of peat, many mines are concentrated in the Mekong Delta with average concentration of C at 17.29% N at 1.2%, P_2O_5 at 0.16%; K_2O at 0.3%; pH: 4.5 and humic acid at 12.8% (**Table 5**). This is a great source of raw materials to supply organic matter to produce organic fertilizer. In addition, seaweed around the coast of Vietnam is a rich source of potassium, micro nutrients or phosphorite ore in many Northern

	C:N ratio	OM	C	N	P_2O_5	K_2O	CaO	MgO	SiO_2 (%)
Rice straw	78-88		54-56	0.64-0.69	0.05-0.11	2.0-2.1	0.42-1.2	0.3-0.52	4.9*
Rice hull	70-106		39-52	0.48-0.70	0.11-0.46	0.28-1.3	0.21-0.34	0.09-0.4	12.7
Rice bran	18-22	67-78	50-55	2.0-2.4	3.60-4.47	1.43-2.45	0.13-0.35	1.11-1.78	
Corn stalks	68		55	0.81	0.37	1.61	0.35	0.48	4.1
Sorghum stalks	73		53	0.73	0.25	1.94	0.60	0.62	3.9
Soybean stems	40		51	1.28	0.14	1.63	0.18	0.15	2.9
Peanut stems	30		42	1.30	0.37	1.31	1.97	1.15	2.5
Peanut hull	28		49	1.73	0.37	1.27	1.96	0.77	1.8
Coconut shell	37	96	53	1.43	0.18	0.50	0.36	0.20	

Table 2.
Composition of some crop residue (source: Wang et al. [13]).

	C:N ratio	C (%)	N (%)	P_2O_5 (%)	K_2O (%)	CaO (%)	MgO (%)	Cu ---mg/kg---	Zn
Cattle	19-28	25-40	0.89-2.1	0.55-4.81	1.6-3.5	0.20-2.0	0.83-2.1	20	122
Goat	16-21	36-48	1.6-2.4	1.5-5.27	1.9-4.0	1.3-5.4	0.7-1.40		
Swine	17-31	4-54	1.6-2.9	1.0-7.1	0.16-1.93	0.8-9.0	0.15-1.7	510	624
Egg chicken	9-14	27-32	0.6-2.9	1.4-6.8	0.77-3.8	0.73-8.2	0.3-1.8		
Meat chicken	11-28	25-47	1.8-2.5	2.11-6.6	1.41-3.6	1.57-21	0.5-1.5	80	724

Table 3.
Composition of some animal waste (source: Wang et al. [13]).

Concentration (%)	Head	Backbone	Tail	Oar
Protein	42.68	37.91	30.36	37.23
Lipid	28.79	37.91	45.10	41.57
Ash	23.13	20.11	15.24	18.47
Carbohydrate	5.4	4.57	2.76	3.70

Table 4.
Chemical composition of catfish processing byproduct (source: Trinh [15]).

provinces such as Thanh Hoa, Hoa Binh, Thai Nguyen, Bac Can, Lang Son and Cao Bang is an additional source of phosphorus and nutrient elements in the organic fertilizer production process.

4.3 Organic fertilizer production and state management

According the Department of Plant Protection of MARD, until December 2017, in Vietnam the number of organic fertilizers including organic mineral fertilizer and bio-organic fertilizer produced and traded were 713, accounting for 5% of the total fertilizer products. There were 180 companies permitted for producing organic fertilizer in Vietnam with the total production capacity of 2.5 million tons/year, accounting for 8.5% of total fertilizer production capacity in whole country. The demand of organic fertilizer is approximately 6 million tons/year and will be increase in the future [10]. In the period 2015–2017, Vietnam exported organic fertilizer to 34 different countries with the export volume in 2017 approximately 76,000 tons, up more than six times compared to 2015 (12,000 tons). In 2015, there were 17 organic fertilizer products exported, in 2016 increased to 56 products and in 2017 there were a total of 75 organic fertilizer products exported abroad. In 2015, there were only two organic fertilizer exporters, in 2016 there were 12 enterprises, by 2017 there were 19 enterprises participating in exporting organic fertilizers.

By the end of 2016, there were 24 Vietnamese standards (TCVN) issued related to organic fertilizer, which focused mainly on testing methods to determine the density and biological activity of microorganisms in the compost enrichment inoculants and content of limiting factors in organic fertilizer. Basically, the standard system of fertilizers in general and organic fertilizer in particular has been built since the 1990s of the previous century, but still lacks in quantity, quality and unresponsive practical requirements. Some additional biological substances in fertilizers such as amino acids, vitamins, plant growth regulators, etc. do not have standard methods for testing and controlling fertilizer quality. Some standards, especially standards related to bio-organic fertilizer have not been reviewed,

Parameters	Mining area			
	North Vietnam	Central South Vietnam	Cuulong delta	Average
pH KCl	3.47 (2.40–6.40)	4.12 (3.74–4.58)	3.95 (3.18–4.78)	3.97 (2.40–6.40)
OC (%)	19.80 (8.58–43.08)	21.81 (16.45–26.54)	29.75 (10.71–40.69)	22.55 (8.58–43.08)
OM (%)	43.56	47.98	65.45	49.61
Total N (%)	0.45 (0.20–0.72)	1.35 (0.88–1.91)	0.96 (0.34–1.54)	1.12 (0.20–1.91)
Total P_2O_5 (%)	0.054 (0.03–0.09)	0.162 (0.08–1.50)	0.062 (0.02–0.13)	0.141 (0.02–1.50)
Total K_2O (%)	0.039 (0.02–0.06)	0.136 (0.10–0.20)	0.652 (0.33–2.26)	0.191 (0.02–2.26)

Table 5.
Composition of peat in Vietnam (source: Hien [14]).

updated, modified to suit the reality and development of production and use of organic fertilizers.

Currently in Vietnam, there are 12 permitted laboratories for testing of quality criteria, limiting factors in organic fertilizers and compost enrichment inoculant. In general, the testing laboratories have met the quality control requirements for general fertilizers and organic fertilizers in particular. However, there are still many issues that need to be considered to improve the effectiveness of fertilizer quality control.

Over the years, Vietnam has achieved certain results in the development of production and use of organic fertilizers. Besides the achieved results, the organic fertilizer industry still faces many difficulties and limitations to be able to develop effectively and sustainably, namely:

a. Farmers are now using inorganic fertilizers because of effectiveness, but not paying attention on the long-term harms of inorganic fertilizer abuse such as soil degradation, environmental pollution, toxic residues in agricultural products, etc.

b. The number of producer of inorganic fertilizer at the present is many times higher than organic fertilizers, which is one of the causes of serious imbalance in production and use of fertilizers.

c. Production technology of organic fertilizer is low with simple and old equipment resulting in low performance and efficiency.

d. There are no specific policies to encourage production and use of organic fertilizers.

e. Agricultural extension programs to introduce and promote the use of organic fertilizers have not been given adequate attention. There are no specific programs of the state or enterprises to support farmer to use organic fertilizers.

f. The set of standards for fertilizer quality control is still incomplete, so it still faces many difficulties in the quality management and registration of organic fertilizer.

5. Vietnam policy on the organic fertilizer production and application

In 2018 ministry of agriculture and development setting up the program to encourage the organic fertilizer production and application with the aim to develop the organic fertilizer contributing to promoting crop production in the direction of enhancing added value and protecting the environment. It concentrates on the followings:

- Effective using the agricultural by-products to produce organic fertilizer for domestic consumption and export;
- Increasing the organic fertilizer at least 3 million tons/year for domestic consumption and export of 0.5 million tons/year;
- Selection and adoption of advanced technology of organic fertilizers production in accordance with the Vietnam conditions;
- Increase the ratio of organic fertilizer products to total fertilizer products from 5% currently to 10% in the near future;
- Encouraging and mobilizing to ensure at least 50% of the fertilizer producer in the country commit to invest in the development of organic fertilizer production and complete the standards, the testing laboratory in service of state management on the organic fertilizer quality control and insurance.

The solution to carry out the program of encouragement of the organic fertilizer production and application is determined as follows:

1. The government should review the legal documents on fertilizer and organic fertilizer to create a suitable legal system for state management of organic fertilizer, including specific contents on further encouraging the production and use of organic fertilizer in the Law on crops production and supporting policies on land use, taxes, credit as well as promotion of application new technologies. In the long term, it is necessary to setting up the priority policies to encourage the production and use of organic fertilizers using available materials from crop production, animal husbandry, food processing waste and other natural material like peat, seaweed etc.

2. The government will develop a national plan on fertilizer production based on the balance between inorganic and organic fertilizers to pushing the gradually increase of proportion of production and use of organic fertilizers. In nearly future it needs to implement a survey projects on the production and use of organic fertilizers for each region in whole country, with special emphasis on local available materials, the feasibility of transferring advanced production technologies, practices of using organic fertilizers, etc. to have a scientific basis to develop a strategy for developing organic fertilizer.

3. Based on the results of reviewing the system of standards of fertilizer, the responsible ministries should speeding up the amendment, supplement and completion of standards for organic fertilizer supporting the quality control and quality assurance (QC&QA) of organic fertilizer. In addition, the testing laboratory system should be reviewed, evaluated and step by step upgraded to

meet the requirement as prescribed by law. The government will invest for the fertilizer testing laboratories in the North, Central and South regions for the quality control and quality ensurance of fertilizer and organic fertilizer.

4. Building up and implementation of new policies to encourage and develop the link chain in crop production and organic fertilizer production, application as well as encourage the organizations and individuals to invest in research, technology transfer, mastering and application of advanced technologies for organic fertilizer production based on Vietnam's available raw materials.

5. Vietnam will promote the research, transfer and application of advanced technologies for organic fertilizer production with priority on environmentally friendly technologies and technologies using locally available materials and tools as well as technology increasing the efficiency use of organic fertilizer, etc. contributing to increase sustainable crop productivity.

6. Regarding inspection and state management of fertilizers quality control and quality assurance (QA&QC), the government will innovating the inspection, examination and compliance with regulations on fertilizer management in all stages from laboratory testing, field trial, production, trading and using fertilizers. The Ministry of Agriculture and Rural Development will organize an specialized inspection forces to check the quality of fertilizers produced and commercialized, thoroughly handling fertilizer producer that fail to meet the conditions for fertilizer production or that have the products not been permitted for the commercialization. The responsible local authorities should be strengthened in the inspection, supervision of production, business and fertilizer use.

7. The government will develop the training materials for management agencies, organizations and individuals producing, trading organic fertilizers as well as organize the training course on the implementation of legal documents and management skills for responsible local authorities. The agricultural extension will innovate the guide on using organic fertilizer through practical models and field days in coordinating local authorities, fertilizer producer to guide the farmer to use organic fertilizer.

8. Regarding the communication, the mass media will coordinate with fertilizer associations, Farmers' Association, Gardening Association, universities, research institutes, etc. propagating and guiding the farmer to produce traditional organic fertilizers based on reuse of agricultural byproducts, animal manure and household waste as well as propagating and replicating advanced models in production, business and use of organic fertilizer. The farmer should understand the role and long-term effects of the use of organic fertilizers via communication.

9. Vietnam government encourages and promotes the international cooperation on organic fertilizer development in Vietnam and will actively participate in the international organic fertilizer market. The international cooperation in the technology transfer in organic fertilizer production from oversee will be strengthened. Vietnam will participate in international treaties and agreements on organic agriculture and organic fertilizer, both multilateral and bilateral with countries and organizations in the region and the world.

6. Conclusion

Vietnam is a tropical country and has enormous progress and remarkable growth in agriculture contributing actively in poverty reduction, national food security, and social stability in last 30 year. Vietnam faces bright opportunities in both domestic and international markets; yet effectively competing in these will depend upon the ability of farmers and firms to deliver products with reliability, and with assurances relating to quality, safety, and sustainability. Organic agriculture using organic fertilizer is one of Vietnam government priority. Vietnam has good condition for organic fertilizer production and application, but the production capacity is small not meet the demand for organic agriculture. Vietnam government promotes the organic fertilizer production and application and has the policy to develop the organic fertilizer in Vietnam.

Author details

Pham Van Toan[1*], Ngo Duc Minh[1] and Dao Van Thong[2]

1 Vietnam of Academy Agricultural Sciences, Hanoi, Vietnam

2 Institute for Agricultural Environment, Hanoi, Vietnam

*Address all correspondence to: toanvaas@gmail.com

References

[1] Ministry of Agriculture and Rural Development. The 2018 summary report of the agricultural sector. 2018

[2] General Statistical Office. Statistical Yearbook 2016. Hanoi, Vietnam: Statistical Publishing House; 2017

[3] Vietnam Customs. Preliminary assessment of Vietnam international merchandise trade performance in whole year 2018. 2018

[4] Tin D. Fertilizer Industry Report. FPTS (FPT Securities joint-stock company). Available from: http://www.fpts.com.vn/FileStore2/File/2015/08/11/FPTSFertilizer%20Industry%20Report.2015.pdf

[5] Mordor Intelligence. Vietnam Organic Fertilizers Market (2016-2021), a Sample Report 2017. 2017

[6] Khanh D, Thanh NH. Management of agricultural waste and potential for co-benefits. In: Presentation presented at the Regional Workshop of the Greater Mekong Sub region on the National Strategy of Integrated Solid Waste Management/3R; Hai Phong; 2010

[7] Tung TQ. Assessment of Policies and Its Implementation on Use of Plant Protection Products in Vietnamese Agriculture. Hanoi: Vietnam Union of Science and Technology Associations; 2015

[8] Willer H. Yussefi M, Lernoud J. The world of organic agriculture - Statistics and Emerging Trends 2018. 2018. https://www.ifoam.bio/en/news/2018/06/.../download-world-organic-agriculture-2018-fr...

[9] Hung LV. Status of organic agriculture development in Vietnam. In: Report Presented at the International Forum Vietnam Organic Agriculture Development and Integration; December 15-16; Hanoi; 2017

[10] Department of plant protection-MARD. Report on reality and solution for development of organic fertilizer. In: Proceeding of Conference on the Development of Organic Fertilizer in Vietnam; Hanoi; 2018. pp. 1-64

[11] Misra R, Roy R, Hiraoka H. On Farm Composting Methods. Rome: Food and Agriculture Organization of United Nations; 2003

[12] Toan PV. Beneficial microorganisms used for organic and biofertilizer. In: Proceeding of Conference on the Development of Organic Fertilizer in Vietnam; Ha Noi; 2018. pp. 81-94

[13] Wang C-H, Lin Y-W, Huang W-T, Chiu L-R. Raw Materials Used for Composting. Available from: http://www.fftc.agnet.org/library.php?func=view&id=20110804115924&type_id=2, 2005

[14] Hien BH. Organic fertilizer from production to utilization. In: Proceeding of Conference on the Development of Organic Fertilizer in Vietnam; Hanoi; 2018. pp. 95-110

[15] Trinh MV. Environmental problem regarding the management of agricultural byproducts used for organic fertilize. In: Proceeding of conference on the development of organic fertilizer in Vietnam Hanoi; 2018. pp. 111-132

[16] Quynh HT, Kazuto S. Organic fertilizers in Vietnam's markets: Nutrient composition and efficacy of their application. Sustainability. 2018;**10**:2437. DOI: 10.3390/su10072437